Ocular Drug Delivery: Advances, Challenges and Applications

Richard T. Addo
Editor

Ocular Drug Delivery: Advances, Challenges and Applications

Springer

Editor
Richard T. Addo
School of Pharmacy
Union University
Jackson, TN
USA

ISBN 978-3-319-47689-6 ISBN 978-3-319-47691-9 (eBook)
DOI 10.1007/978-3-319-47691-9

Library of Congress Control Number: 2016954017

Printed on acid-free paper

This Springer imprint is published by the registered company Springer International Publishing AG part of Springer Nature
The registered company address is: Gewerbestrasse 11, 6330 Cham, Switzerland

To my wife, children and to God Almighty

Preface

Most people will agree with me that one of their greatest fears is losing their eyesight. Eye diseases affect the quality of life of hundreds of millions of individuals around the world. Scientists continue their search for the ideal ocular drug delivery system but they are still being faced with challenges due to the complex nature of the eye. This global observation and my research into ocular drug delivery gave birth to this book "Ocular Drug Delivery: Advances, Challenges, and Applications".

This book has detailed chapters that are comprised of various topics and issues pertinent to ocular therapeutic and drug delivery systems. It describes the anatomy of the eye with the barriers affecting the delivery of drug to it. The physiological barriers to drug delivery and the targeted delivery to specific compartments of the eye have also been discussed. Topical and systemic drug delivery to the eye has been compared. The delivery of proteins and peptides into the ocular cavity through novel delivery systems such as nano- and microparticles, nano micelles, and others that have been used to deliver drugs especially to the posterior segment of the eye has also been discussed with a detailed method of evaluating these novel systems in vivo, in vitro, and ex vivo. With all these novel drug delivery systems, scientists are still researching advanced drug delivery systems, some of which are described in this book.

My appreciation to all the authors for their outstanding job, and their dedication in contributing to this book, and Springer Protocols for their encouragement and patiently working with all of us and making this dream come through.

Jackson, USA　　　　Richard T. Addo

Acknowledgments

I would like to thank the Almighty God for His divine strength, guidance, wisdom, and faithfulness throughout this journey. Thanks to all the distinguished authors for their invaluable contributions and cooperation. It has been a delightful experience working with all the individual contributors of the chapters and their desire in sharing their knowledge, concepts, and expertise in various aspects of ocular drug delivery.

I am grateful to friends, former and current colleagues at University of Science & Technology, Mercer University, Shenandoah University, George Washington University, and Union University for their mentor and encouragement.

My sincere appreciation to my wife, Evelyn, for her unfailing love, my children, Tetteybea, Richard Jr, Jesse David, for their contagious smiles and my parents, Salome and Michael, for their continual inspiration.

Finally, I would like to acknowledge the support of my publisher, Springer, for making this book a reality.

Contents

Contributors

Evelyn Addo School of Social Work, College of Education & Human Studies, Union University, Jackson, TN, USA

Richard T. Addo Department of Pharmaceutical Sciences, School of Pharmacy, Union University, Jackson, TN, USA

Javed Ahmad Department of Pharmaceutics, National Institute of Pharmaceutical Education and Research (NIPER), Raebareli, Uttar Pradesh, India

Sohail Akhter LE STUDIUM® Loire Valley Institute for Advanced Studies, Centre-Val de Loire, France; Nucleic Acids Transfer by Non-Viral Methods, Centre de Biophysique Moléculaire, Orléans, France

Oluyemisi A. Bamiro Department of Pharmaceutical Sciences, School of Pharmacy, Union University, Jackson, TN, USA; Department of Pharmaceutics and Pharmaceutical Technology, Olabisi Onabanjo University, Ago Iwoye, Nigeria

Lunawati Bennett School of Pharmacy, Union University, Jackson, TN, USA

Yvonne Darko School of Pharmacy, College of Health Sciences, University of Ghana, Accra, Ghana

Henry Nettey School of Pharmacy, College of Health Sciences, University of Ghana, Accra, Ghana

Rodney Siwale Department of Pharmaceutical Sciences, College of Pharmacy, Western New England College of Pharmacy, Springfield, MA, USA

Ruhi V. Ubale Department of Pharmaceutical Sciences, School of Pharmacy, Union University, Jackson, TN, USA; School of Pharmacy, Lake Erie College of Osteopathic Medicine, Bradenton, Florida, USA

Ameeduzzafar Zafar Department of Pharmaceutics, Faculty of Pharmacy, Jamia Hamdard University, New Delhi, India

Chapter 1
Background of Ocular Drug Delivery

Oluyemisi A. Bamiro, Ruhi V. Ubale and Richard T. Addo

Abstract Delivery of drugs to the eye is one of the challenges facing clinicians due to the unique anatomy, physiology, and biochemistry of the eye. Current treatment protocols for administration of drugs in eye diseases are primarily solution, gels or ointments. However, these modes of delivery have several drawbacks such as short residence time, short duration of action, the need for repeated administrations and non-specific toxicity. This chapter will introduce a brief discussion on various topics and issues relevant to ocular therapeutics and drug delivery.

Keywords Ocular · Eye · Drug delivery · Formulations

Introduction

The eye is a highly complex organ with a unique anatomy and physiology that protects it and even makes drug delivery to the eye a major challenge to formulation scientists. The eye is divided into anterior and posterior segments (Chap. 2 of this book discusses the detailed anatomy of the eye). Glaucoma, conjunctivitis, blepharitis and cataract are some of the diseases that affect the anterior segment of the eye, while age-related macular degeneration (AMD) and diabetic retinopathy affect the posterior segment. The topical route is the most convenient and patient compliant form of administering drugs to the anterior segment of the eye. Eye drops

O.A. Bamiro (✉) · R.V. Ubale · R.T. Addo
Department of Pharmaceutical Sciences, School of Pharmacy,
Union University, Jackson, TN 38305, USA
e-mail: Bamroy67@yahoo.co.uk

O.A. Bamiro
Department of Pharmaceutics and Pharmaceutical Technology,
Olabisi Onabanjo University, Ago Iwoye, Nigeria

R.V. Ubale
School of Pharmacy, Lake Erie College of Osteopathic Medicine,
Bradenton, Florida, USA

R.T. Addo (ed.), *Ocular Drug Delivery: Advances, Challenges and Applications*,
DOI 10.1007/978-3-319-47691-9_1

accounts for about 90 % of drugs administered through this route, less than 5 % of the drug administered gets absorbed due to washing off of drugs by tear dilution, reflex blinking, nasolachrymal drainage and other ocular barriers (static and dynamic) in the eye (Gaudana et al. 2010a; Patel et al. 2013). Due to the problems mentioned above, administration of drugs to the posterior segment of the eye using the conventional eye drop does not achieve therapeutic drug concentration.

Intravitreal and periocular routes are recommended for delivery of drugs to the posterior segment of the eye. These routes have their own limitations: repeated injections through intravitreal route is painful, thus reducing patient compliance. In periocular route, there is ease of administration but the static and dynamic barriers constitutes a problem.

In view of all these, novel formulations have been investigated for the delivery of drugs especially to the posterior segment of the eye to overcome the ocular barriers. Some of these formulations include: sols, liposomes, dendrimers, microparticles, nanoparticles, nanomicelles, implants, contact lenses, microneedles and in situ thermosensitive gels.

Ocular Barriers

There are two major barriers in the eye: blood-aqueous barrier (BAB) and blood-retina barrier (BRB).

BAB consists of the endothelium of the iris/ciliary blood vessels and the non-pigmented ciliary epithelium, both located in the anterior segment of the eye. These cells form tight junctions of the "leaky" type. BRB consists of two types of cells viz retinal capillary endothelia cells (inner BRB) and retinal pigment epithelium cells (RPE) (outer BRB). Both have tight junctions of the "non-leaky" type thus restricting movement of drugs into the retina. This explains why after systemic administration of drugs, the therapeutic level is hardly achieved in the retina (Gaudana et al. 2010b; Occhiutto et al. 2012). These barriers are discussed in detail in Chap. 3 of this book.

Drug Delivery to the Eye

The division of the eye into anterior and posterior segment is actually adapted for the ocular delivery of drug since the method and route of administration of a drug is dependent on the part of the eye segment (Cohen 2012).

Topical instillations are efficient when delivering drugs to the anterior segment of the eye. Despite the relatively small proportion of administered drug dose that reaches the tissues in the anterior segment of the eye, topical instillation still remains effective because of the high concentrations of drug administered (Geroski and Edelhauser 2000). Preservatives such as benzalkonium chloride present in eye drops have been shown to act as permeability enhancers thus improving the penetration of drugs into ocular tissues (Chen et al. 2011). For years, delivery of therapeutic doses to the posterior segment tissues posed a significant challenge to ophthalmologists. Systemic, intraocular and periocular routes were been used but the challenges posed by the ocular barriers were not allowing therapeutic doses to be delivered to the targeted tissues.

In the early 2000s, liposomal verteporfin was injected systemically coupled to laser infrared therapy to close abnormal choroidal neovessels complicating AMD (Wormald et al. 2005; Cohen 2012). A few years later, anti-vascular epithelia growth factor (anti-VEGF) was developed for the treatment of wet AMD using repeated intravitreous injections (Vedula and Kryzstolik 2008), thus marking the beginning of a new era for delivery of drugs into ocular tissues.

New Trends in Ocular Drug Delivery

Prodrug Design

Prodrugs, developed about six decades ago, are derivatives of drug molecules that are chemically or enzymatically transformed in vivo into the active parent drug (Stella and Himmelstein 1980). Prodrugs are designed to improve solubility, stability, bioavailability and also to decrease side effects of ophthalmic drugs. For successful utilization of prodrugs, recognition of drug properties and participation of barriers at target site are critical factors. Prodrugs are useful for improving drug permeation by enhancing lipophilicity and by modulating aqueous solubility. Different approaches are used:

Functional Group Prodrug Approach Ester prodrugs are derived from carboxylic or hydroxyl functional groups. These functional groups exist in ionized form in physiological condition thereby limiting the passage of drug through the lipid membrane leading to reduced bioavailability. The first ophthalmic prodrug to be produced was dipivefrine, an ester prodrug of epinephrine to decrease intraocular pressure (Hussain and Truelove 1976; Jarvinen and Jarvinen 1996). Corneal penetration with dipivefrine was 10 times better than epinephrine (Wei et al. 1978; Mandell et al. 1978).

Transporter Target Prodrug Approach Prodrugs are synthesized in such a way that they will have an affinity for influx transporters, which sees them as substrates and thus transports them across the membrane. Examples of transporters used in ocular delivery include peptides (Anand et al. 2003; Janoria et al. 2010), amino acids (Balakrishnan 2002; Katragadda 2008) and in recent times vitamins such as biotin (Janoria et al. 2009) and vitamin C (Dalpiaz 2005) are being used.

Lipid Prodrug This approach was designed to improve lipophilicity of hydrophilic drugs molecules hence increasing corneal permeability. The drug molecule is covalently bound to a lipid moiety such as diglyceride or phosphoglyceride, fatty acids. Lipid prodrugs diffuse across a cell membrane by facilitated diffusion thereby resulting in improved cellular absorption. If the drug is too lipophilic, permeability will be limited.

Dendrimers

They are macromolecular "tree-like" nanostructured polymer with several reactive end group that forms an internal core. The unique structure makes them solubilize poorly water-soluble drugs (Cheng et al. 2008) and also mimic globular proteins (Esfand and Tomalia 2001; Hecht and Frechet 2001). Dendrimers have been reported to enhance corneal residence time of topically administered drugs (Yang et al. 2012), target neuroinflammation in retinal degeneration (Iezzi et al. 2012), used as sutures upon removal of cataract (Wathier et al. 2004).

Liposomes

They are vesicular systems comprising of an aqueous inner core and an outer layer made up of phospholipid bilayers from a natural or synthetic origin. They are classified based on lipid bilayer and sizes such as SUVs (small unilamellar vesicle), MLV (multilamellar vesicle), LUV (large unilamellar vesicle) and GUV (giant unilamellar vesicle). Liposomes can be used to deliver both hydrophilic and hydrophobic drugs to the eye, which is one of the advantages apart from being biocompatible and non-toxic (Schwendener 2007; Mishra et al. 2011). However, liposomes have limitations such as stability, sterilization, and low encapsulation efficiency. They can become chemically unstable due to hydrolysis or oxidation of their unsaturated lipids or physically unstable due to leakage of drugs which could lead to aggregation (Agarwal et al. 2016). Rathode and Deshpande suggested that the stability problems might be overcome by lyophilization (Rathode and Deshpande 2010). Coating of liposomes with high molecular weight chitosan has been shown to reduce aggregation, increase viscosity thus improving residence time in the cornea (Mehanna et al. 2010).

Nanomicelles

They are self-assembled colloidal structures in the size range of about 10–100 μm consisting of an inner hydrophobic core and an outer hydrophilic shell (Trivedi and Kompella 2010). Nanomicelles are divided into three categories: Polymeric, Surfactants and Polyionic complex micelles (PIC). Surfactant micelles have high critical micelle concentration (CMC) and are unstable in solution whereas, polymeric micelles have low CMC and are stable on dilution (Vaishya et al. 2014). PIC are water soluble with a charged core-forming block. The driving force for micelle formation is the electrostatic charge interaction between this block and the oppositely charged drug stabilizes the micelle. Selection of carrier depends on the physicochemical properties of the drug, site of action, the interaction between drug and carrier and biocompatibility (Vaishya et al. 2014). The shape and size of surfactant micelles depend on the concentration of surfactants, pH, temperature and the ionic strength of the surfactant. Nanomicelles have the following advantages: minimal drug degradation, easy permeation through the ocular epithelium, enhanced bioavailability and no irritation (Cholkar et al. 2012).

Implants

They are polymeric devices inserted into the eye surgically, to prolong the release of a drug in the eye. Polymers used could be biodegradable or non-biodegradable. The non-biodegradable polymers include polyvinyl alcohol (PVA), ethylene vinyl acetate etc. have been used for sustained release of implants such as Retisert®, I-vation® and Iluvien®. The problem with them is that after the drug has been depleted, the polymer carrier remains in the eye and it needs to be removed surgically and this can lead to non-compliance. If left in the eye, accumulation could lead to other problems like cataract, hemorrhage, retinal detachment and other complications.

The biodegradable polymers that have been used in ocular drug delivery include but not limited to poly lactide-co-glycolide acid (PLGA) and Polylactic acid (PLA). These polymers degrade in the eye when the drug has been depleted leaving no residue. Examples of implants using biodegradable polymers are Ozurdex® and Surodex® (Haghjou et al. 2011).

Contact Lens

They are curved shaped discs prepared from silicon containing hydrogels or polyvinyl alcohol hydrogels inserted into the eye to cover the cornea. It is separated from the cornea by a thin fluid layer called the postlens tear film. The postlens tear film allows drugs released from the lens to have a residence time of at least 30 min in front of the cornea since the fluid in tear film does not mix readily with a tear in the eye (Creech et al. 2001). Drug bioavailability was increased from less than 5 % in eye drops to about 50 % in contact lenses.

In an experiment conducted by Guzman-Aranguez et al. (2012), they loaded drugs into the lenses by soaking in drug solutions. However this method has its limitations: adequate drug was not loaded and also the drug loaded was released within a few hours. Thus, soaked contact lenses cannot provide slow extended drug release.

Nanoparticles, microparticles, and liposomes can also be loaded on matrix of contact lens to deliver drug to the eye (Gulsen and Chauhan 2004, 2005)

Another method is known as molecular imprinting. The drug (acting as a template) is added before the polymerization of the polymer units. The monomers will arrange as a function of their ability to interact with drug molecules. After polymerization, the drug molecule is removed. This imprinting makes it easier to load more drugs and also to extend the release of the drug (Hirantani et al. 2005; Alvarez-Lorenzo et al. 2006; Tieppo et al. 2012).

The supercritical solvent method is another way of loading drugs onto contact lenses. The drug is dissolved in a high volatile fluid such as carbon dioxide which is brought near the matrix to load the drug (Braga et al. 2012).

Microneedle

They are individual minute needles or array of micrometer-sized needles that are manufactured by adapting the tools of the microelectronics industry (Jiang et al. 2007). They were originally developed for transdermal drug delivery. Microneedles which are solid or hollow were made from stainless sheet metals (Jiang et al. 2007) while glass microneedles were made from borosilicate micropipette tubes (Cho and Olsen 2013). Drugs coated on microneedles were delivered in a minimally invasive manner via interscleral and intracorneal routes (Jiang et al. 2007).

Fenestrated microneedles were developed by Khandan and co-workers and they found that more drugs could be loaded onto this microneedle. These fenestrated microneedles can also be used to load nanoparticles and microparticles (Khandan et al. 2015).

In Situ Thermosensitive Gels

They are polymeric aqueous solutions that change from a sol to gel in response to temperature changes. An example of polymer in this group is poloxamer. Polymers that are thermosensitive behaves as liquid below its low critical solution temperature (LCST). When environmental temperature reaches LCST or above, it forms a gel (Cao et al. 2007).

In situ thermosensitive gels have been shown to increase residence time and also to control release of drugs in the eye (Sultana et al. 2003).

References

Agarwal R, Iezhitsa I, Agarwal P, Nasir NAA, Razali N, Alyautdin R, Ismail NM (2016) Liposomes in topical ophthalmic drug delivery: an update. Drug Deliv 23(4):1075–1091

Alvarez-Lorenzo C, Yanez F, Barreiro-Iglesias R, Concheiro A (2006) Imprinted soft contact lenses as norfloxacin delivery systems. J Control Release 113:236–244

Anand BS, Nashed YE, Mitra AK (2003) Novel dipeptide prodrugs of acyclovir for ocular herpes infections: Bioreversion, antiviral activity and transport across rabbit cornea. Curr Eye Res 26 (3–4):151–163 [PubMed: 12815543]

Balakrishnan A, Jain-Vakkalagadda B, Yang C, Pal D, Mitra AK (2002) Carrier-mediated uptake of L- tyrosine and its competitive inhibition by model tyrosine linked compounds in a rabbit corneal cell line (SIRC)–strategy for the design of transporter/receptor targeted prodrugs. Int J Pharm 247(1–2):115–125

Braga ME, Costa VP, Pereira MJ, Fiadeiro PT, Gomes AP, Duarte CM, de Sousa HC (2011) Effects of operational conditions on the supercritical solvent impregnation of acetazolamide in Balafilcon A commercial contact lenses. Int J Pharm 420:231–243

Cao Y, Zhang C, Shen W, Cheng Z, Yu L, Ping Q (2007) Poly (N-isopropyl acrylamide)-chitosan as thermosensitive in situ gel forming system for ocular drug delivery. J Controlled Release 120:186–194

Chen W, Li Z, Hu J et al (2011) Corneal alternations induced by topical application of benzalkonium chloride in rabbit. PLoS ONE 6(10):e26103

Cheng YY, Xu ZH, Ma ML, Xu TW (2008) Dendrimers as drug carriers: applications in different routes of drug administration. J Pharm Sci 97(1):123–143

Cho SW, Olsen TW (2013) Drug delivery to the suprachoroidal space In: Ocular drug delivery systems: barriers and application of nanoparticulate systems. CRC Press Taylor and Francis Group, pp 241–242

Cholkar K, Patel A, Vadlapudi AD, Mitra AK (2012) Novel nano micellar formulation approaches for anterior and posterior segment ocular drug delivery. Recent Patents Nanomedicine 2(2):82–95

Cohen-Behar F (2012) Drug delivery to the eye: current trends and future perspectives. Therapeutic Delivery 3(10):1135–1137

Creech JL, Chauhan A, Radke CJ (2001) Dispersive mixing in the posterior tear film under a soft contact lens. Ind Eng Chem Res 40:3015–3026

Dalpiaz A, PavanS B, Scaglianti M, Vitali F, Bortolotti F, Biondi C, Scalturin A, Manfredini S (2005) Vitamin C and 6-amino-vitamin C conjugates of diclofenac: synthesis and evaluation. Int J Pharm 291(1–2):171–181 [PubMed: 15707744]

Esfand R, Tomalia DA (2001) Poly (amidoamine) (pamam) dendrimers: from biomimicry to drug delivery and biomedical applications. Drug Discov Today 6(8):427–436

Gaudana R, Ananthula HK, Parenky A, Mitra AK (2010a) Ocular drug delivery. AAPS J. 12:348–360. doi:10.1208/s12248-010-9183-3

Gaudana R, Ananthula HK, Parenky A, Mitra AK (2010b) Ocular drug delivery. The AAPS Journal 12(3):348–360

Geroski DH, Edelhauser HF (2000) Drug delivery for posterior segment eye disease. Invest Ophthalmol Vis Sci 41(5):961–964

Gulsen D, Chauhan A (2004) Ophthalmic drug delivery through contact lenses. Invest Ophthalmol Vis Sci 45:2342–2347

Gulsen D, Chauhan A (2005) Dispersion of microemulsion drops in HEMA hydrogel: a potential ophthalmic drug delivery vehicle. Int J Pharm 292:95–117

Guzman-Aranguez A, Colligris B, Pintor J (2012) Contact lenses: promising devices for ocular drug delivery. J Ocul Pharmacol Ther 29(2):189–199. Available on https://www.researchgate.net/publication/233873735

Haghjou N, Soheilian M, Abdekhodaie MJ (2011) Sustained release intraocular drug delivery devices for treatment of uveitis. J Ophthalmic Vis Res 6(4):317–329

Hecht S, Frechet JMJ (2001) Dendritic encapsulation of function: applying nature's site isolation principle from biomimetics to materials science. Angew Chem Int Ed 40(1):74–91

Hiratani H, Mizutani Y, Alvarez-Lorenzo C (2005) Controlling drug release from imprinted hydrogels by modifying the characteristics of the imprinted cavities. Macromol Biosci 5:728–733

Hussain A, Truelove JE (1976) Prodrug approaches to enhancement of physicochemical properties of drugs IV: novel epinephrine prodrug. J Pharm Sci 65(10):1510–1512

Iezzi R, Guru BR, Glybina IV, Mishra MK, Kennedy A, Kannan RM (2012) Dendrimer-based targeted intravitreal therapy for sustained attenuation of neuroinflammation in retinal degeneration. Biomaterials 33(3):979–988

Janoria KG, Boddu SH, Wang Z, Paturi DK, Samanta S, Pal D, Mitra AK (2009) Vitreal pharmacokinetic of biotinylated ganciclovir: role of sodium-dependent multivitamin transporter expressed on retina. J Ocul Pharmacal Ther 25(1):39–49

Janoria KG, Boddu SH, Natesan S, Mitra AK (2010) Vitreal pharmacokinetics of peptide-transporter- targeted prodrugs of ganciclovir in conscious animals. J Ocul Pharmacal Ther 26(3):265–271

Jarvinen T, Jarvinen K (1996) Review: Prodrugs for improved ocular drug delivery. Adv Drug Deliv Rev 19:203–224

Jiang J, Gill HS, Ghate D, McCarey BE, Patel SR, Edelhauser HF, Prausnitz MR (2007) Invest Ophthalmol Vis Sci 48(9):4038–4043

Katragadda S, Jain R, Kwatra D, Hariharan S, Mitra AK (2008) Pharmacokinetics of amino acid ester prodrugs of acyclovir after oral administration: interaction with the transporters on Caco-2 cells. Int J Pharm 362(1–2):93–101

Khandan O, Kahook MK, Rao MP (2015) Fenestrated microneedles for ocular drug delivery. Sens Actuators B 223:15–23

Mandell AI, Stentz F, Kitabchi AE (1978) Dipivalyl epinephrine: a new pro-drug in the treatment of glaucoma. Ophthalmology 85(3):268–275

Mehanna MM, Elmaradny HA, Samaha MW (2010) Mucoadhesive liposomes as ocular delivery system: physical, microbiological, and in vivo assessment. Drug Dev Ind Pharm 36:108–118

Mishra GP, Bagui M, Tamboli V, Mitra AK (2011) Recent applications of liposomes in ophthalmic drug delivery. J Drug Delivery, 2011, Article ID 863734. doi:10.1155/2011/863734

Occhiutto ML, Freitas FR, Maranhao RC, Costa VP (2012) Breakdown of the blood-ocular barrier as a strategy for the systemic use of nanosystems. Pharmaceutics 4:252–275. doi:10.3390/pharmaceutics4020252

Patel A, Cholkar K, Agrahari V, Mitra AK (2013) Ocular drug delivery systems: an overview. World J Pharmacol 2(2):47–64. doi:10.5497/wjp.v2.i2.47

Rathod S, Deshpande SG (2010) Design and evaluation of liposomal formulation of pilocarpine nitrate. Indian J Pharma Sci 72:155–160

Schwendener RA (2007) Liposomes in biology and medicine. Adv Exp Med Biol 620:117–128

Wormald R, Evans J, Smeeth L, Henshaw K (2005) Photodynamic therapy for neovascular age-related macular degeneration. Cochrane Database Syst Rev 19(4):CD002030

Stella VI, Himmelstein KI (1980) Prodrug and site-specific drug delivery. J Med Chem 23:1275–1282

Sultana Y, Zafar S, Ali A (2003) Enhanced ocular bioavailability with sol to gel system of pefloxacin mesylate in-vitro and in-vivo studies. J Sci Pharm 4(1):5–10

Tieppo A, Pate KM, Byrne ME (2012) In vitro controlled release of an anti-inflammatory from daily disposable therapeutic contact lenses under physiological ocular tear flow. Eur J Pharm Biopharm 81:170–177

Trivedi R, Kompella UB (2010) Nanomicellar formulations for sustained drug delivery: strategies and underlying principles. Nanomedicine (Lond.) 5(3):485–505

Vaishya RD, Khurana V, Patel S, Mitra AK (2014) Controlled ocular drug delivery with nanomicelles. Wiley Interdiscip Rev Nanomed Nanobiotechnol 6(5):422–437. doi:10.1002/wnan.1272

Vedula SS, Krzystolik MG (2008) Antiangiogenic therapy with anti-vascular endothelial growth factor modalities for neovascular age-related macular degeneration. Cochrane Database Syst Rev 16(2):CD005139

Wathier M, Jung PJ, Carnahan MA, Kim T, Grinstaff MW (2004) Dendritic macromers as in situ polymerizing biomaterials for securing cataract incisions. J Am Chem Soc 126(40):12744–12745

Wei CP, Anderson JA, Leopold I (1978) Ocular absorption and metabolism of topically applied epinephrine and a dipivalyl ester of epinephrine. Invest Ophthalmol Vis Sci 17(4):315–321

Yang H, Tyagi P, Kadam RS, Holden CA, Kompella UB (2012) Hybrid dendrimer hydrogel/PLGA nanoparticle platform sustains drug delivery for one week and antiglaucoma effects for four days following one-time topical administration. ACS nano 6(9):7595–7606

Chapter 2
Anatomy of the Eye and Common Diseases Affecting the Eye

Evelyn Addo, Oluyemisi A. Bamiro and Rodney Siwale

Abstract The human eye is a unique part of the human body that has been described as the window to the human soul. The eye being a complex organ is broadly divided into two segments—anterior and posterior segments. Each of these broad segments has specific disease conditions associated with them. This chapter will extensively describe these various components of the eye and the common disease conditions associated with it.

Keywords Anatomy · Physiology · Eye diseases

Introduction

The human eye is one of the most complex parts of the body. Among the sense organs in the body, the eye is the most widely used. Most of the information we gather about our surroundings are things we see with our eyes. The eye has been compared to a camera that gathers light and turns it into images.

The eye (Fig. 2.1) is basically divided into two segments: (1) the anterior segment that consists of the cornea, iris, pupil, conjunctiva, ciliary body, anterior

E. Addo (✉)
School of Social Work, College of Education & Human Studies, Union University, Jackson, TN, USA
e-mail: evelynoaddo@gmail.com

O.A. Bamiro
Union University School of Pharmacy, Jackson, TN, USA

O.A. Bamiro
Department. Pharmaceutics and Pharmaceutical Technology, Olabisi Onabanjo University, Ago Iwoye, Ogun, Nigeria

R. Siwale
Department of Pharmaceutical Sciences, College of Pharmacy, Western New England College of Pharmacy, Springfield, MA, USA

R.T. Addo (ed.), *Ocular Drug Delivery: Advances, Challenges and Applications*,
DOI 10.1007/978-3-319-47691-9_2

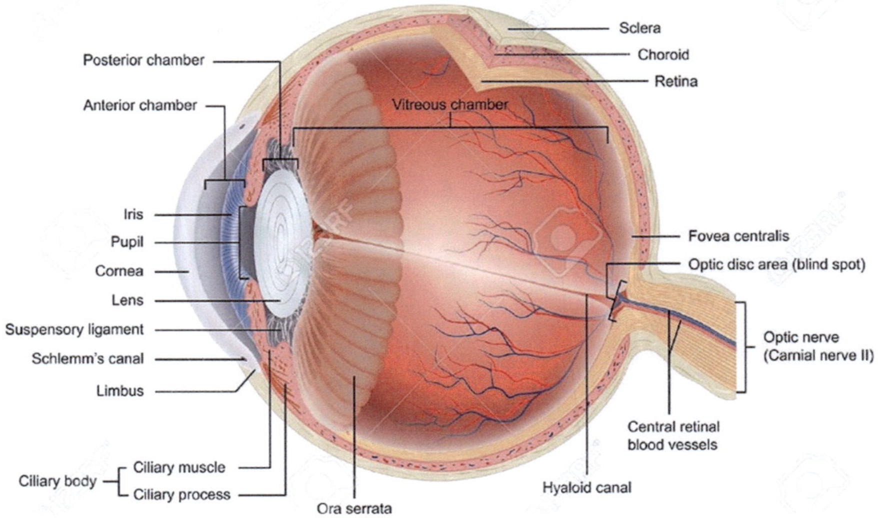

Fig. 2.1 Anatomy of the eye Adapted from Newsplus cited on 24th June 2016 available at http://www.newsplus.mobi/the-eyes/

chamber, aqueous humor, trabecular meshwork, and lens. (2) The posterior segment consists of vitreous humor, sclera, choroid, retina, macula and optic nerve.

Anterior Segment

Cornea

It is the outermost part of the eye. It is clear with no blood vessels to nourish it but it obtains its nourishments from the tear and aqueous humor. It is an organized group of cells and proteins arranged in five layers (Vaishya et al. 2014). It provides two-thirds of the total ocular refractory power.

The tear film bathes the corneal surface and helps protect the eye from irritants.

Epithelium

It is the outermost layer of the cornea, about 50 μm thick. It is composed of a layer of basal cells, 4–5 layers of non-keratinized stratified squamous epithelia cells held together by tight junctions to form a barrier against fluid loss and penetration of foreign materials such as dust, bacteria, etc. The epithelium is filled with tiny nerve endings that make the cornea very sensitive to pain. The basement membrane is the part of the epithelium that serves as the foundation on which the epithelial cells anchor and stay organized (National eye Institute).

Bowman's Layer

It lies directly below the basement membrane of the epithelium, a transparent sheet of tissue composed of strong-layered protein fibers called collagen. The collagen is finer than those in the corneal stroma. The compact arrangement of the collagen fibers in this layer makes it resistant to trauma. Once destroyed, it will not regenerate. It helps the cornea to maintain its shape.

Stroma

Comprises about 90 % of cornea's thickness (500 μm), the majority (78 %) of which is made up of water and the remaining is collagen that gives strength, elasticity and form to the cornea. Collagen fibrils are arranged in flat bundles called lamellae and are arranged at right angles to each other. Stroma collagen fibrils compose of type I and V collagen that forms complex surrounded by proteoglycan consisting of keratin sulfate or chondroitin sulfate/dermatan sulfate side chain. They help in the regulation of hydration and structural properties of the cornea (DelMonte and Kim 2011).

Descemet's Membrane

The Descemet's membrane is a thin but strong sheet of tissue that serves as a protective barrier against infection and injuries. It is composed of collagen fibers made by the endothelial cells that lie below it. It is regenerated readily after injury. It thickens throughout life because it is constantly being produced, hence making it resistant to enzymatic degradation and toxins. They are about 3 μm at birth and gets up to 10 μm with age.

Endothelium

The innermost layer of the cornea that keeps it clear. It is on the posterior surface of the cornea facing the anterior chamber. It serves two functions to maintain the health and clarity of the stroma. It pumps excess fluid that leaks into the stroma out and allows nutrients and molecules from the aqueous humor In addition to these, they are responsible for the synthesis of the Descemet's membrane (Yu et al. 2011).

Iris

It is a circular, thin structure in the eye responsible for controlling the pupil. It consists of two layers: pigmented fibrovascular layer known as stroma and the second layer is pigmented epithelia cells. The iris separates the space between the lens and the cornea into two chambers: the anterior chamber (between cornea and iris) and the posterior chamber (between iris and lens). The color of the iris is established genetically and it depends on the amount of pigment in the iris.

Pupil

It is an opening in the iris through which light passes before reaching the lens to be focused on the retina. The size of the pupil depends on the amount of light it is exposed to and the muscles of the iris control the size (Albert and Gamm 2016).

Conjunctiva

It is a thin membrane that lines the inside of the eyelid and covers the sclera. It is composed of non-keratinized stratified epithelium and goblet cells. It helps to protect the eyes by producing mucus that helps in lubricating the eyes and prevents entry of microorganisms. The conjunctiva firmly adheres to the edge of the cornea and there are loose folds in the cul-de-sac that allow for easy movement of the eye. The conjunctiva is highly vascularized; this explains why the eye is always red in response to injury or disease (Agarwal et al. 2016). The conjunctiva consists of three parts: the palpebral, which lines the under surface or posterior surface of the eyelid, bulbar and fornix (Vision RX 2005). The palpebral is subdivided into three: the marginal conjunctivae, tarsal conjunctivae and orbit conjunctivae. The bulbar is the thinnest part of the conjunctiva. It is so transparent that the underlying white sclera and vessels are seen clearly. The fornix is the fold lining the cul-de-sac formed by conjunctiva covering the posterior surface of the lids to the conjunctiva covering the anterior surface of the globe. The goblet cells are unicellular glands which secret mucin. They are round and oval and may produce up to 2.2 ml of mucous daily to lubricate and protect epithelial cells. It also reduces the surface tension of tear film thus ensuring its stability (Majumder 2008).

Ciliary Body

It is a ring-shaped thickened tissue inside the eye, consisting of ciliary muscle that helps the lens in focusing properly and ciliary epithelium that helps in the secretion of aqueous humor. It is joined to the lens by connective tissues known as zonular fibers. The ciliary body forms part of the uvea. The anterior portion is known as pars plicata characterized by the ciliary process while posterior part is pars plana, which has a relatively flat and pigmented inner surface (Borges-Giampani and Junior 2013). The ciliary muscles consists of three muscle fibers:

Longitudinal or Meridional Fibers

These are the external part that attaches the ciliary body at the anterior part to the scleral spur and trabecular meshwork at the limbus, and at the posterior to the supra-choroidal lamina fibers that connect choroid and sclera (Bruce 1997). Contraction of these fibers leads to the opening of the trabecular meshwork and Schlemm's canal.

Oblique or Radial Fibers

These are in the middle connecting longitudinal and circular fibers. They originate in the sclera spur and are attached to collagenous substance near the ciliary process. The contraction of these fibers may widen the uveal trabecular spaces.

Circular or Sphincter Fibers

These run parallel to the limbus. Contraction of these fibers makes the zonules relax, increasing the lens axial diameter and its convexity.

Anterior Chamber

It is the part of the eye that contains the aqueous humor ($\sim$0.25 mL), about 3 mm deep.

Aqueous Humor

Aqueous humor is a clear fluid that fills and helps form the anterior and posterior chambers of the eye. It has been stated earlier that the aqueous humor provides nutrition for the cornea which is avascular structure that must remain clear to allow light transmission, and therefore cannot be invested within a vasculature. The aqueous humor is analogous to a blood surrogate for these avascular structures. It removes excretory products from metabolism, transports neurotransmitters, stabilizes the ocular structure and contributes to the regulation of the homeostasis of these ocular tissues. Aqueous humor also permits inflammatory cells and mediators to circulate in the eye in pathological conditions, as well as drugs to be distributed to different ocular structures (Sires 1997). Aqueous humor is an important component of the eye's optical system secreted by the ciliary epithelium. The major components of the aqueous humor are organic and inorganic ions, carbohydrates, glutathione, urea, amino acids, proteins, oxygen, carbon dioxide and water. It is slightly hypertonic to plasma in a number of mammalian species (Goel et al. 2010).

Trabecular Meshwork

It consists of connective tissues surrounded by endothelium, divided into three parts (Van Buskirk 1986; Llobet et al. 2003):

Uveal Meshwork

This is formed by prolongations of connective tissue arising from the iris and ciliary body stroma totally covered by endothelial cells. Uveal meshwork forms the lateral border of the anterior chamber, extending from the iris root and ciliary body to the peripheral cornea, consists of bands of connective tissue, with irregular openings that measure between 25 and 75 μm (Goel et al. 2010).

Corneoscleral Meshwork

This is characterized by the presence of lamellae covered by endothelium–like cells standing on a basal membrane. The lamellae formed by glycoproteins, collagen, hyaluronic acid and elastic fibers. It extends from the scleral spur to the anterior wall of the scleral sulcus and is the most extensive portion of the trabecular meshwork. It is composed of perforated sheets that become progressively smaller nearing Schlemm's canal (Goel et al. 2010). The corneoscleral meshwork has four layers: connective tissue with collagen fiber layer, elastic fiber layer, "glass membrane" layer (delicate filaments embedded in ground substance) and endothelia layer (Gong et al. 1989; Goel et al. 2010).

Juxtacanalicular Meshwork

This is formed by cells embedded in a dense extracellular matrix, is in direct contact with the inner wall of endothelial cells from Schlemm's canal, comprised of endothelial cells surrounded by connective tissue like a vein. Schlemm's canal possesses internal collector channels and is connected to episcleral and conjunctival veins through the external collector channels, the intrascleral venous plexus, the deep scleral plexus and the aqueous veins (Gong et al. 1996; Goel et al. 2010).

Lens

It is the clear and elastic part behind the pupil. It can easily change its shape due to its elasticity. The lens loses its elasticity and ability to focus light sharply with age (mostly 40–50 years). At times, the lens can become hardened, clouded and difficult to focus a condition known as cataract.

Posterior Segment

Vitreous

A transparent, colorless gel liquid that fills the space between the lens and the retina. Vitreous contains about 98 % water, hyaluronic acid and collagen fibrils that form a scaffolding and there is absence of blood vessels. The fluid is stagnant if substances enter the humor it has to be removed surgically because it can affect a person's field of vision.

Sclera

It is the white part of the eye, opaque and elastic, consisting of collagen fibers. The opacity is due to the irregular arrangement of the collagen fibers. Its function is to protect and maintain the globe shape of the eye. It is thicker in males than females, thickens gradually towards the cornea. The sclera is divided into three parts:

Episclera

This is the outermost part of the sclera, composed of fibers of collagen, loose fibrous and elastic tissues that attach to the Tenon capsule. There are three types of blood vessels in the episclera:

- Bulbar conjunctival plexus: a superficial plexus of fine vessels (arteries) overlying and freely moveable over the episclera.
- Episcleral plexus: straight radially arranged vessels (veins) in the superficial episclera. They are also moveable over the deep layers (although less easily than the bulbar conjunctival vessels).
- Deep episcleral plexus (also called scleral plexus): a crisscross of vessels closely applied to the sclera.

Sclera Stroma

This contains irregularly arranged collagen bundles, proteoglycans, glycoproteins, and fibroblasts, which play an important role in the synthesis of proteoglycans and glycoproteins.

Lamina Fusca

This is the innermost layer of the sclera containing pigmented cells or melanocytes that migrate from the choroid. The suprachoroidal space separates it from the choroid.

Choroid

It is a vascular layer of the eye containing connective tissues, fibroblasts, melanocytes and resident immunocompetent cells that lie between the sclera and the retina (Nickla and Wallman 2010). The choroid is the posterior portion of the uveal tract (ciliary body, iris, and choroid) which supplies oxygen and nourishment to the retina. The smooth inner surface is attached to the retinal-pigmented epithelium (RPE), while the outer surface is attached to the sclera at the optic nerves and exit at vortex veins. The choroid helps in light absorption thus reducing the amount of light that enters the retina (in species with pigmented choroids), thermoregulation via heat dissipation, and modulation of intraocular pressure (IOP) via vasomotor control of blood flow. The choroid also plays an important role in the drainage of the aqueous humor from the anterior chamber, via the uveoscleral pathway (Nickla and Wallman 2010). Its thickness decreases from 200 μm at birth to about 80 μm by age 90 (Ramrattan et al. 1994).

Generally, the choroid is divided into four layers:

- Haller's layer is the outermost layer and consists of posterior ciliary arteries and large lumen veins
- Sattler's layer contains intermediate size blood vessels
- Choriocapillary layers consist of capillaries
- Bruch's Membrane is the innermost layer.

Retina

It is a circular disc of about 30–40 mm in diameter (Kolb 1991), and 0.5 mm thick. The retina is the light-sensitive area at the back of the eye consisting of millions of cells tightly packed together. It is divided into three basic groups based on the cell type: photoreceptor cells, neuronal cells, and glial cells.

Photoreceptor Cells

Photoreceptor cells consist of cones and rods. The rods are narrower than the cones, highly sensitive to light, work in dim light, while cones function only in bright light (Baylor et al. 1979). There are three types of cones based on level of response to short, medium or long wavelength thus leading to color separation. Each rod and cone contain a photoreceptor element and an axon. Each photoreceptor consists of an outer segment containing a photon-capturing photo pigment (opsin) and an inner segment containing mitochondria and adenosine triphosphate (ATP).

Neuronal Cells

Neuronal consist of bipolar, ganglion, horizontal and amacrine cells. The bipolar cells connect the photoreceptors to the ganglion cells that has dendrites synapse with bipolar cells. The horizontal cells that are responsible for allowing the eyes to adjust to seeing well connect the bipolar cells together. Neurons present in the outer plexiform layer of the retinal are also interconnected by the horizontal cells. Amacrine cells connect the bipolar and ganglion cells with each other, there are about forty types of amacrine cells. It functions within the inner plexiform layer.

Glial Cells

Glial cells also are known as supporting cells (Muller cells, astrocytes, and microglial cells) are interspersed between and among the axons in the ganglion cells of the retina and the optic nerve. Muller cells, the principal glial cells of the retina, form a supporting matrix radially across the thickness of the retina and form the inner and outer limiting membranes of the retina. Muller cell bodies sit in the inner nuclear layer and project irregularly thick and thin processes in either direction to the outer limiting membrane and to the inner limiting membrane. Muller cell, found between cell bodies of the neurons in the nuclear layers and envelope groups of neural processes in the plexiform layers. Retinal neural processes allow direct contact at their synapses.

The retina consists of several different layers (Fig. 2.2a, b). Listed below are the layers from inside the vitreous outwards to the choroid:

- The inner limiting membrane is the boundary between retina and vitreous
- The nerve fiber layer contains axons of ganglion nuclei

(a)

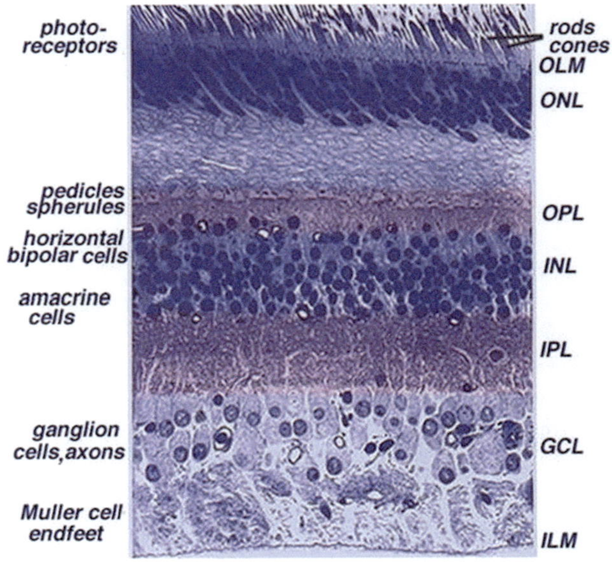

Light micrograph of a vertical section through the human retina.

OLM = Outer limiting membrane (separates inner segments of photoreceptors from their cell nuclei)
ONL = Outer nuclear layer (cell bodies of the retinal rods and cones.)
INL = Inner nuclear layer (nuclei of horizontal, bipolar and amacrine cells.)
IPL = Inner plexiform layer (synapses between ganglion cells and bipolar cells)
GCL = Ganglion cell layer
ILM = Inner limiting membrane (boundary between the retina and the vitreous body.)

(b)

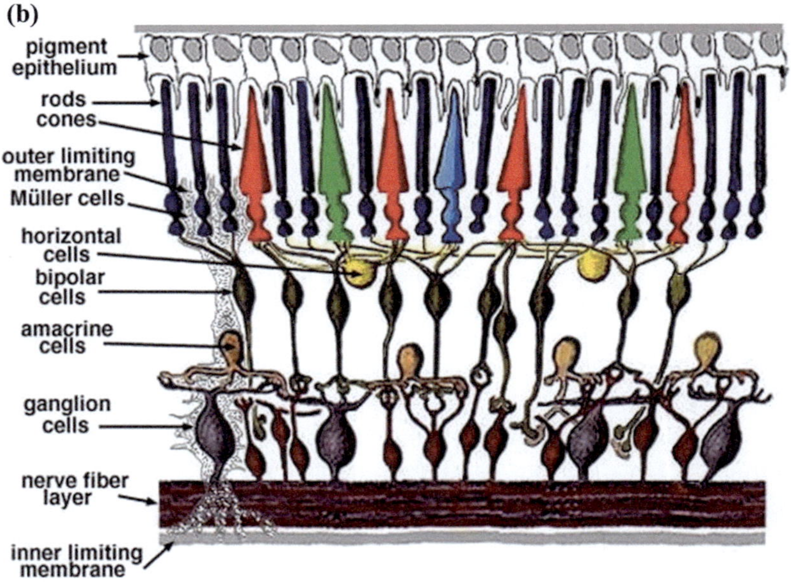

◀ **Fig. 2.2** **a** Light micrograph of a vertical section through the human retina. Adapted from Kolb H, Fernandez E, Nelson R, eds. The organization of the retina and visual system. Salt Lake City, UT: University of Utah Health Sciences Center; 1995. **b** Simple organization of the retina. Adapted from Kolb H, Fernandez E, Nelson R, ed. The organization of the retina and visual system. Salt Lake City, UT: University of Utah Health Sciences Center; 1995. Cited on 10th July 2016. Available on http://webvision.med.utah.edu/book/part-i-foundations/simple-anatomy-of-the-retina/

- The ganglion cells layer contains nuclei of ganglion cells, axons of which become the optic nerve with some displaced amacrine cells
- The inner plexiform layer contains the synapse between the bipolar cell axons and the dendrites of the ganglion and amacrine cells
- The inner nuclear layer contains the nuclei of horizontal, bipolar and amacrine cells
- The outer plexiform layer also known as the outer synaptic layer contains projections of rods and cones axons ending in the rod spherule and cone pedicle. Synapses of these with horizontal cells dendrites and bipolar cell dendrites occur in the macula region known as fiber of Henle
- The outer nuclear layer consists of cell bodies of rods and cones
- The outer limiting membrane is the layer that separates the inner segment portions of the photoreceptors from their cell nuclei
- The rod and cone layer contains the rods and cones photoreceptor cells
- The pigment epithelium is a single layer of cuboidal cells close to the choroid. It contains melanin that helps in absorption of light thus decreasing the intensity

The macula is an oval shaped pigmented area in the retina. The center of which is the fovea, which contains photoreceptor cones that make the fovea responsible for sharp vision necessary for driving and reading. Parafovea (inner) and perifovea (outer) belts surround the fovea (Iwasaki and Inomata 1986).

Optic Nerve

The optic nerve also known as cranial nerve connects the eye to the brain. The optic nerve carries the impulses formed by the retina, the nerve layer that lines the back of the eye, senses light, and creates impulses. It contains about 1.2 million nerve fibers, transfers visual impulses from the retina via electrical impulses to the visual center in the brain. The optic disc also known as blind spot due to absence of photoreceptor cells is the part of the retina where the optic nerve leaves the eye.

Common Diseases Affecting the Eye

Different types of diseases affect the eye and if not treated on time could eventually lead to blindness.

Dry Eye

Dry eye disease also known as keratoconjunctivitis sicca occurs when the eye does not produce tear or the consistency is not clear and it dries up quickly. Dysfunction of the Meibomian gland could also be associated with people having dry eye syndrome. Some of the symptoms are grittiness, ocular dryness, mucoid discharge, hyperemia and photophobia (Foster 2015).

Meibomian Gland Dysfunction (MGD)

Meibomian gland dysfunction (MGD) is blockage of the meibomian glands so they are unable to secrete enough oil into the tears, leading to fast evaporation of the tears. MGD is a leading cause of dry eye syndrome. It is associated with an eyelid problem called blepharitis. The symptoms of MGD are also similar to that of dry eye syndrome—red eyes, grittiness, itchy eyes and blurred vision.

Cataract

Cataract is clouding of the lens in the eye. This disease causes loss of vision in age 40 and above. It is the most common cause of blindness in the world. Clouding can occur at the back of the lens in people with diabetes or those taking steroids. Some of the symptoms of cataracts are cloudy or blurred vision, poor night vision, colors looking faded and multiple images.

Macular Degeneration

Macular degeneration leads to loss of vision and it is age related. It is also known as age-related macular degeneration (AMD or ARMD). AMD is a common eye condition among the leading cause of blindness in the world affecting over 25 million people worldwide, 8 million of whom have severe blindness. It steadily destroys the macula, which is the part of the eye responsible for sharp, central vision. In some patients, AMD can advance rather slowly such that the loss of vision may not occur for a very long time. However, in others, it progresses rapidly leading to complete loss of vision. This disease makes it difficult to recognize faces, drive a car, read and other activities that require central vision (Jager et al. 2008; Coleman et al. 2008). Macular degeneration may be hereditary however smoking, high blood pressure, high cholesterol, obesity, and being light skinned, female, and having a light eye color are also risked factors for macular degeneration (National Eye Institute). There are two types of AMD, dry AMD, and wet AMD. Exudative (wet) AMD is more threatening than the dry type and is responsible for 90 % of cases of severe visual loss in elderly people (Jager et al. 2008).

Vision loss in AMD occurs because of damage to the macula, specifically to the photoreceptor cells (Beatty et al. 2000). Since the original study by Noell et al. showing that exposure of freely moving rats to continuous bright visible (green) light leads to selective degeneration of rod photoreceptor cells, there has been a growing concern that long-term exposure to sunlight may be a contributing factor to the development of AMD (Noell et al. 1966).

There is currently no cure for AMD, however, scientists are consistently working towards developing therapies that can slow the disease or even restore vision.

Hypertensive and Diabetic Retinopathy

Hypertensive and diabetic retinopathy occurs when there is damage to the tiny blood vessels supplying the retina. This could eventually lead to blindness.

Glaucoma

Glaucoma is associated with pressure build up in the eye leading to damage of the optic nerve and loss of vision. The normal eye pressure is known as intraocular pressure (IOP) is below 21 mmHg. There are different types: open-angle glaucoma, closed-angle glaucoma, normal tension glaucoma. Open-angle glaucoma occurs when there is slow drainage of the aqueous humor through the trabecular meshwork, leading to build-up of fluid in the eye, thus causing increase IOP. There is usually no symptom initially except gradual vision loss.

Closed-angle glaucoma occurs when the iris blocks the trabecular meshwork. This is a serious type of glaucoma that can lead to blindness. Some of the symptoms of acute attack are severe eye pain, sudden blurry vision, headache, nausea, and vomiting.

Normal tension glaucoma is when a person has IOP consistently below 21 mmHg, but there is still damage to the optical nerve and loss of vision.

References

Agarwal R, Iezhitsa I, Agarwal P, Nasir NAA, Razali N, Alyautdin R, Ismail NM (2016) Liposomes in topical ophthalmic drug delivery: an update. Drug Deliv 23(4):1075–1091

Albert DM, Gamm DM (2016) Encyclopaedia britannica available on https://www.Britannica.com/science/pupil-eye. Cited 28 May 2016

Baylor DA, Lamb TD, Yau KW (1979) Responses of retinal rods to single photons. J Physiol 288:613–634

Beatty S, Koh H, Phil M, Henson D, Boulton M (2000) The role of oxidative stress in the pathogenesis of age-related macular degeneration. Surv Ophthalmol 45:115–134

Borges-Giampani AS, Junior JG (2013). Anatomy of ciliary body, ciliary processes, anterior chamber angle and collector vessels. doi:10.5772/52780

Bruce Shields M. (1997) Aqueous humor dynamics: anatomy and Physiology. In Textbook of glaucoma, 4th edn. Williams & Wilkins, pp 5–31

Coleman HR, Chan CC, Ferris FL, Chew EY (2008) Age-related macular degeneration. Lancet 372:1835–1845

DelMonte DW, Kim T (2011) Anatomy and physiology of the cornea. J Cataract Refract Surg 37:588–598. Available on https://www.researchgate.net/…/49848146. Retrieved 18 May 2016

Foster CS (2015) Dry eye syndrome (Medscape Homepage. On the internet). Available on http://www.emedicine.medscape.com/article/1210417-overview. Updated 15 July 2015, cited 6 June 2016

Goel M, Picciani RG, Lee RK, Bhattacharya SK (2010) Aqueous humor dynamics: a review. Open Ophthalmol J 4:52–59

Gong H, Tripathi RC, Tripathi BJ (1996) Morphology of the aqueous outflow pathway. Microsc Res Tech 33(4):336–367

Gong HY, Trinkaus-Randall V, Freddo TF (1989) Ultrastructural immunocytochemical localization of elastin in normal human trabecular meshwork. Curr Eye Res 8(10):1071–1082

Iwasaki M, Inomara H (1986) Relation between superficial capillaries and foveal structures in the human retina. Invest Ophthalmol Vis Sci 27:1698–1705

Jager RD, Mieler WF, Miller JW (2008) Age-related macular degeneration. N Engl J Med 358:2606–2617

Kolb H (1991) The neural organization of the human retina. In: Heckenlively JR, Arden GB (eds) Principles and practices of clinical electrophysiology of vision. Mosby Year Book Inc, St. Louis, pp 25–52

Llobet A, Gasull X, Gual A (2003) Understanding trabecular meshwork physiology: a key to the control of intraocular pressure? News Physiol Sci. 18. Available on http://Physiologyonline.physiology.org/content/18/5/205. Cited 2 June 2016

Majumder PD (2008). Anatomy of conjunctiva (eophtha homepage on the internet). Available on www.eophtha.com/eophtha/anatomy/anatomyofconjunctiva.html. Updated 27 July 2008, cited 17 May 2016

Nationaleyeinstitute. Available http://www.nei.nih.gov/amd/summary.asp

Nickla DL, Wallman J (2010) The multifunctional choroid. Prog Retin Eye Res 29(2):144–168

Noell WK, Walker VS, Kang BS, Berman S (1966) Retinal damage by light in rats. Invest Ophthalmol 5:450–473

Ramrattan RS, Van der Schaft TL, Mooy CM, De Bruijn WC, Mulder PGH, De Jong P (1994) Morphometric analysis of Bruch's membrane, the choriocapillaris and the choroid in aging. Invest Ophthalmol Vis Sci 35:2857–2864

Sires B (1997) Orbital and ocular anatomy. In: Wright Ed (ed) Textbook of ophthalmology. Williams and Wilkins, Baltimore, MD

Vaishya RD, Khurana V, Patel S, Mitra AK (2014) Controlled ocular drug delivery with nanomicelles. Wiley Interdiscip Rev Nanomed Nanobiotechnol 6(5):422–437. doi:10.1002/wnan.1272

Van Buskirk EM (1986) Clinical Atlas of Glaucoma. WB Saunders, Philadelphia

Vision RX Encyclopaedia (2005) (Homepage on the internet) Conjunctivae. www.visionrx.com/library/enc/enc_conjuctiva.asp. Cited 18 May 2016

Yu WY, Sheridan C, Grierson I, Mason S, Kearns V, Lo ACY, Wong D (2011). Progenitors for the corneal endothelium and trabecular meshwork: a potential source for personalized stem cell therapy in corneal endothelial diseases and glaucoma. J Biomed Biotechnol. Article ID 412743. doi:10.1155/2011/412743

Chapter 3
Ocular Barriers

Henry Nettey, Yvonne Darko, Oluyemisi A. Bamiro and Richard T. Addo

Abstract Drugs for ocular administration are administered topically and their absorption is limited by different barriers known as ocular barriers. These barriers are located in different areas of the eye and the barrier encountered by a particular drug will depend on the route of administration of the drug to the eye, or the segment of the eye the drug is eliciting its action. Some of these barriers include tear film, the blood-ocular barrier which consists of blood-aqueous barrier and blood-retinal barrier and tight junctions present in the cornea that prevents passage of some drug molecules. This chapter will discuss the physiological barriers to ocular drug delivery which limit the distribution of the drug within the eye and also reduce its bioavailability.

Keywords Ocular barrier · Drug absorption · Blood-retina barrier · Blood-ocular barrier

Introduction

The human eye, a quasi-spherical organ for vision, is a fluid filled organ enclosed with layers of tissue and composed of various muscles that enable it to function.

The first and outer layer of tissue is the sclera. This is a tough, opaque fibrous tissue which together with the internal pressure maintains the spherical shape of the eye. Towards the anterior part of the eye, this tissue is transformed into the cornea. The

H. Nettey (✉) · Y. Darko
School of Pharmacy, College of Health Sciences, University of Ghana, Accra, Ghana
e-mail: hnettey@msn.com

O.A. Bamiro
Union University School of Pharmacy, Jackson, TN, USA

O.A. Bamiro
Pharmaceutics & Pharmaceutical Technology, Olabisi Onabanjo University, Ago Iwoye, Nigeria

R.T. Addo
Department of Pharmaceutical Sciences, School of Pharmacy, Union University, Jackson, TN, USA

R.T. Addo (ed.), *Ocular Drug Delivery: Advances, Challenges and Applications*,
DOI 10.1007/978-3-319-47691-9_3

specialized transparent tissue of the sclera allows light rays into the eye (Grandolfo et al. 1991).

The middle layers of tissue, the uvea, are constituted of the iris, ciliary body, and choroid. The iris is the colored portion of the eye seen through the cornea. It possesses muscles for neurally controlled adjustment of the pupil—an aperture in its center. This action with the pupil moderates the amount of light entering the eye. The ciliary body is a ring of tissue around the eye lens with two components—a muscular component and vascular component. The muscular component adjusts the eye lens for accommodation. On the other hand, the vascular component of the ciliary body, also called the ciliary processes, produces fluid to fill the frontal space between the lens and the cornea. This fluid, the aqueous humor, nourishes the cornea and lens, contributes to the shape of the eye and protects against bacterial infection with its bactericidal properties (Egger et al. 1994). The aqueous humor drains at a junction between the cornea and iris. There it goes through a porous tissue—the trabecular meshwork, into the Schlemm's canal, a collecting channel which empties into the veins and subsequently the systemic circulation (Anon 2013). The choroid is a rich capillary bed and is the main blood supply for the photoreceptors of the retina.

The retina is the final and innermost layer of tissue. It possesses neurons, originating from the brain, sensitive to light and is responsible for transmitting visual signals to the brain for vision (Anon 2013).

Posteriorly the eye is filled with a thick gelatinous substance called the vitreous humor. This fluid makes up to about 80 % of the total volume of the eye and occupies the void between the lens and the surface of the retina. In addition to its structural function of maintaining the shape of the eye, the vitreous humor also eliminates, via phagocytic cells, blood and other particles that might interfere with light transmission to the retina (Purves et al. 2001).

A film of tears rich in mucin produced by the lacrimal glands covers the cornea and keeps it moist. There is also the conjunctiva. This membrane covers the internal part of the eyelid and the anterior portion of the sclera (Grandolfo et al. 1991).

Ocular Barriers

Drugs for ocular activity are administered topically as eye preparations, mostly as eye drops. Ocular drug absorption is limited by barriers which control the uptake of fluids and solutes into the anterior and posterior parts of the eye. These, in turn, protect the eye from the impact of otherwise harmful pharmaceuticals but reduce the bioavailability of ocular drugs. These barriers, known as ocular barriers, comprise of the tear film, cornea, conjunctiva, blood-ocular and blood-retina barriers. The blood-retina barrier is principally against drug absorption to the posterior portions of the eye whereas the rest act in opposition to anterior absorption of ocular drugs.

Tear Film

This tissue is a film of fluid covering the surface of the cornea and contained between the lid margins (Anon 2008). As the primary physiological barrier to topically instilled ocular drugs, it plays a complex and active role in regulating epithelial functions and interacting closely with surrounding tissues of the eye (Barar et al. 2009). Similar to an extracellular matrix, it provides the nutrients and communication pathway, distributes regulatory factors and provides a pathway for cells (e.g., inflammatory cells) as well as is the primary source of protection for the eye against injury either through chemical, mechanical, bacterial or viral means. The tear film is generally described as composed of a complex structure of three main layers—aqueous, lipid and mucous or mucin layer, riding on the hydrophobic surface of the epithelium (Winter et al. 2010). However, there is no clear distinction between the mucous and aqueous layers which in actuality exist as a mucoaqueous layer (Gipson 2004; Benitez-del-Castillo and Lemp 2012; Yokoi et al. 2014).

Lipid Layer

It is formed from the oily product secreted by the meibomian glands, large sebaceous glands located in the tarsal plates of the eyelids, onto the upper and lower lid margins and lies in the outer layer of the film. It consists of an outer non-polar lipid layer and an inner polar lipid layer, with intercalated proteins. The polar layer stabilizes the non-polar layer (Rolando and Zierhut 2001; Nichols et al. 2011; Benitez-del-Castillo and Lemp 2012). The key function of these glandular lipids is to provide an outer hydrophobic barrier to prevent the evaporation of tears and enhance the stability of the tear film (Mishima and Maurice 1961; Bron et al. 2008; Mathers 2004; Nichols et al. 2011). The absence of an outer oily layer leads to an increased evaporation rate of the aqueous contents by about 10–20 times (Mishima and Maurice 1961; Maïssa and Guillon 2010).

The lipid layer is also essential in the formation of sharp visual images as its constitution also provides for light refraction into the eye.

Mucoaqueous Layer

This consists of the mucous and aqueous layers in a mucoaqueous mixture. The inner mucin layer, consisting of both gel-forming and trans-membrane mucins, is produced by conjunctival goblet cells and by conjunctival and corneal epithelial cells. The lacrimal glands under stimulation by hormonal, sympathetic or parasympathetic systems produce the aqueous content of the layer. It serves a variety of purposes such as removal of foreign bodies, cells, toxic materials and pathogens, transfer of nutrients to the cornea, enabling the movement of cellular material over the ocular surface and tear film stabilization (Gipson 2004; Winter et al. 2010; Rolando and Zierhut 2001).

Blood Ocular Barriers

The bioavailability of systemically administered ocular drugs to the posterior segments of the eye, via oral or intravenous routes, is reduced due to poor ocular blood flow and the presence of blood-ocular barriers. This is, also, accompanied by an increase in drug side effects as high doses are to be administered before therapeutic efficacy is achieved. Effective therapeutic doses might be achieved by intravitreal administration of such drugs though drug diffusion of lipophilic and charged compound across the ocular milieu is also hindered by these same blood-ocular barriers. Besides, this route of administration offers problems such as retinal detachment and low compliance (Barar et al. 2009).

The blood-ocular barriers are the blood-aqueous barrier and the blood-retinal barrier. These exert their effect on drug penetration and ocular bioavailability by their composition and positioning in the eye.

Blood-Aqueous Barrier

The Blood-aqueous barrier (BAB) is located in the anterior part of the eye. It is formed by endothelial cells of the blood vessels within the iris as well as the non-pigmented cell layer of the ciliary epithelium. These cell layers possess tight junctions responsible for inhibiting the movement of solutes into the anterior milieu which maintains the chemical equilibrium of the ocular fluids (Freddo 2001).

The BAB inhibits the permeation of macromolecules such as plasma albumin into the aqueous humor (Urtti 2006). Small sized hydrophobic drugs, however, can penetrate and are eliminated rapidly via the uveal circulation. However, a macromolecule such as horse radish peroxidase (HRP, 40 kDa) is able to transverse the fenestrated capillaries of the ciliary body (Hornof et al. 2005; Barar et al. 2009).

The reduction in solute movement across the iris blood vessels into systemic circulation maintains the transparency of the eye. However, inflammation of the BAB may disrupt the integrity of the barrier and consequent unlimited drug distribution to the anterior parts of the eye (Urtti 2006).

Blood-Retinal Barrier

The blood-retinal barrier (BRB) provides restricted permeability between the blood and the retina. It operates on two levels, the retinal vessels and chorioepithelial surface—composed of retinal capillary endothelial (RCE) cells and retinal pigment epithelial (RPE) cells, forming the inner and outer BRB, respectively (Cunha-Vaz 1976).

The retinal pigment epithelium (RPE) monolayer, via the sub retinal space, separates the outer surface of the neural retina from the choroid, sustains and maintains the function of the neural retina. Its functions also include the absorption of light, dispersion of heat energy derived from incident light; and participation in the immune response of the retina (Marmor 1998; Hornof et al. 2005).

The RPE has robust barrier restrictiveness but allows some intrinsically specialized transport processes. These processes modulate passage of nutrients and compounds such as pharmaceutical products and thus enable their selective uptake and interchange between the choroid and retina (Hormof et al. 2005; Duvvuri et al. 2003). RPE cells, particularly polarized ones, regulate intracellular Na^+ and K^+ homeostasis via Na^+, K^+-ATPase on their apical surface (Quinn and Miller 1992). The outer layer of the RPE also possesses tight junctions (zonula occludens) which seal off intercellular spaces and serve as tight barriers.

The inner BRB, composed of retinal capillary endothelial (RCE) cells in the lumen of retinal capillaries also is a barrier to circulating molecules in the blood circulation. They possess a high degree of tight junctions in the endothelial microvasculature of the retina. There is also the presence of Glial cells, such as astrocytes, in such tight junctions within the endothelium which exhibit barrier characteristics similar to the Blood Brain Barrier though with a greater density than in the brain (BBB) (Janzer and Raff 1987; Jaleh and Yadollah 2008; Stewart and Tuor 1994; Cunha-Vaz 1997; Gardner et al. 1999).

The RCE barrier prevents the penetration of proteins and small lipophobic substances though studies have shown it allows the uptake of lipophilic materials (Hornof et al. 2005; Sunkara and Kompella 2003). As such systemic (via oral or intravenous administration) and intravitreal administration of drugs only enable sufficient bioavailability and pharmacological effects of the ocular drugs posteriorly in the eye (Barar et al. 2009).

Although this is possible, the effectiveness of the blood-retina barrier (BRB) inhibits effective ocular penetration of systemically administered drugs. For instance, the bioavailability of systemically administered drugs is about 1–2 % of the drug concentration in the plasma (Duvvuri et al. 2003). This, therefore, requires that such drugs be administered in higher concentrations with consequent increase in their side effect profiles (Selvin 1983).

On the other hand, intravitreal drug administration offers a more straightforward access to the vitreous and retina though drug diffusion, particularly of larger and positively charged drugs, across to the choroid is hindered by the RPE barrier (Urtti 2006). It should be noted, however, that active transport mechanisms or passive diffusion across the BRB eliminates drugs from anterior portions and vitreous humor. As such for drugs administered intravitreally, also, highly lipophilic drugs experience rapid elimination from the vitreous humor across the retinal barrier into systemic circulation. However drugs diffusing initially into the aqueous humor, whose diffusion through the BRB is hindered, would experience a longer vitreous half-life. Such drugs undergo removal, for instance, via diffusion across the iris surface (Maurice and Mishima 1984; Pitkänen et al. 2003).

The Conjunctiva and Cornea—Corneal and Non-corneal Routes of Drug Absorption

The bioavailability of topically administered drugs, as is implemented by most ocular drugs is reduced by lacrimal drainage and systemic absorption from the washing of the conjunctiva (Dartt et al. 2006; Macha and Mitra 2003; Mitra et al. 2006). Nevertheless, small-sized hydrophobic molecules are absorbed via the cornea whereas large hydrophilic ones are absorbed via the conjunctiva and sclera (Macha and Mitra 2003; Ahmed 2003).

The human cornea consists of 5 layers—the epithelium, basement membrane (Bowman's layer), stroma, Descemet's membrane and endothelium. This tissue contains various layers that limit transcorneal permeation of drugs. The outer superficial epithelial cells display tight junction complexes whereas, the wing and basal cells exhibit gap junctions. The stroma and Descemet's membrane cover the inner endothelial cells which contain macula adherents but allow the transverse of materials (Barar et al. 2009). The cornea is an important mechanical and chemical barrier, which limits the access of exogenous substances into the eye and protects the intraocular tissues. The cornea is a clear and avascular structure with average diameter and thickness of 12 mm and 520 mm, respectively (Hornof et al. 2005).

The corneal epithelium has desmosome-attached cells with intercellular gap junctions which allow permeation of small molecules. The surface of the epithelium has tight junctions (zonulae occludens) which also prevent diffusion across its cells, especially macromolecular and hydrophilic molecules. From studies done in rabbits, only relatively small molecules can permeate through the pores, which have an average diameter of 2.0 nm (Hämäläinen et al. 1997). The stroma, however, due to the high percentage of hydrated collagen, is more hydrophilic and rather inhibits the transverse of lipophilic molecules. Charged molecules may also experience inhibition in permeation. The pores of the corneal epithelium are negatively charged at physiological pH. This is due to the presence of negatively charged carboxylic groups of the tight junction proteins (Hornof et al. 2005). As a result, negatively charged molecules due to repulsive forces experience difficulty in permeation through the pores. Ionic interaction between the negatively charged tight junction proteins and positively charged molecules, on the other hand, also pose a hindrance to the permeation of such molecules (Rajasakul and Robinson 1989). In all, the cornea passively allows the transfer of materials across its cells and this transfer is influenced by various factors, such as the lipophilicity, molecular weight, charge, and degree of ionization of the drug (Schoenwald and Huang 1983; Hornof et al. 2005). However, after transcorneal transport, the drugs diffuse into the aqueous humor and then to the anterior uvea but cannot reach the posterior portions of the eye at sufficient therapeutic concentrations (Duvvuri et al. 2003). This is a reduced diffusion of drugs across to the vitreous humor from the anterior portions.

The cornea maintains its transparency by an active process of fluid transport and ion transfer though the endothelium and inflammation affect these mechanisms and in turn its transparency. To address this, however, it possesses to some extent some immune response (Sunkara and Kompella 2003).

The conjunctiva is a mucous membrane consisting of an outer epithelium, which is two to three cell layers thick, and an underlying vascularized connective tissue richly supplied by blood vessels. Its epithelium functions in the maintenance of the tear film by the production of mucus to its surface. With the cornea, it presents a barrier to topically administered ocular drugs from the presence of tight junctions on the apical surface of its cells (Barar et al. 2009). Upon drug administration, a greater proportion of the drug is lost through the systemic circulation.

The first barrier for the permeation of topically applied drugs into the eye via the non-corneal route is the bulbar conjunctiva. This route allows the transverse of large and hydrophilic substances though these are poorly absorbed through the cornea (Ahmed 2003; Ahmed and Patton 1985). Rabbit studies showed that the conjunctival epithelium has 2 times larger pores and 16 times higher pore density than the corneal epithelium resulting in a 15 to 25-fold higher permeability of molecules in comparison to the cornea (Hämäläinen et al. 1997).

Due to the nature of the conjunctiva, the bioavailability of drugs is greatly reduced as they permeate the conjunctival/scleral pathway into the anterior portions of the eye (Urtti and Salminen 1993; Ahmed 2003).

The sclera, the next ocular tissue, however, enables the diffusion of materials as it is composed mainly of collagen and polysaccharides and relatively poorly vascularized than the conjunctiva. It allows for 10 times permeation than the cornea and half as permeable as the conjunctiva (Hämäläinen et al. 1997).

Drug Elimination

Drug elimination from the aqueous humor occurs via two mechanisms: by aqueous turnover through the chamber angle and Schlemm's canal and by the venous blood flow of the anterior uvea (Maurice and Mishima 1984). The elimination via the chamber angle and the Schlemm's canal is dependent on the convective flow and independent of the drug characteristics. Uveal elimination, however, depends on the

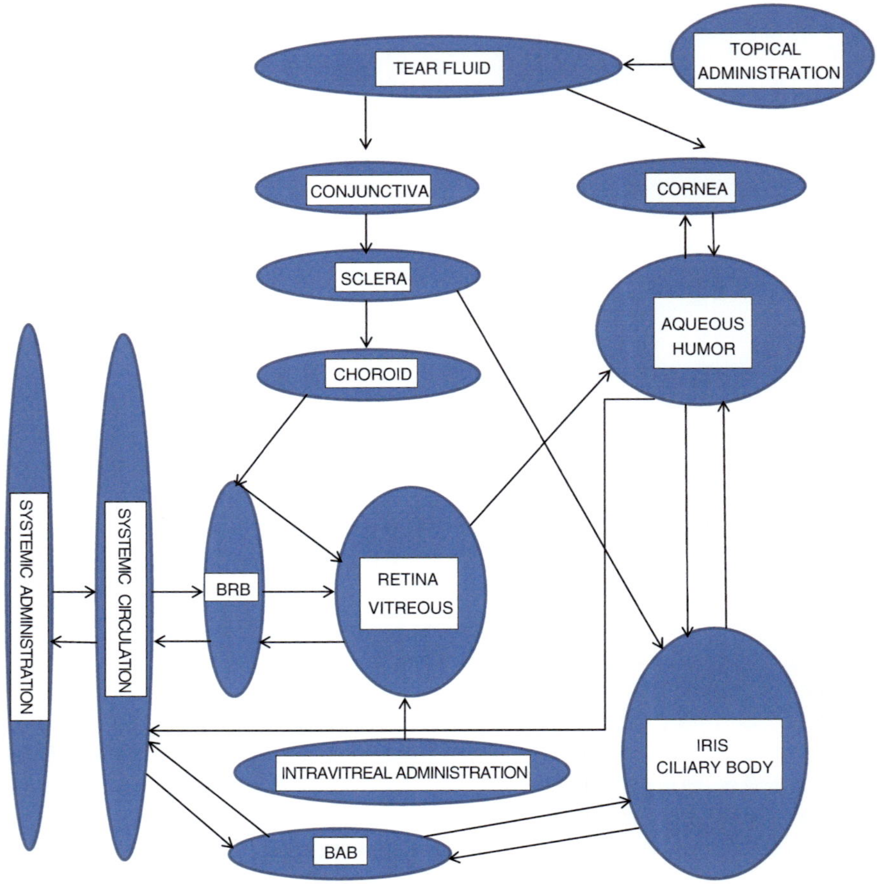

Fig. 3.1 Ocular penetration routes for drugs after topical, systemic and intravitreal administration. The ocular barriers (*BRB*, blood-retinal barrier; *BAB*, blood-aqueous barrier, tear film, conjunctiva and cornea) adapted from Hornof et al. (2005)

extent of lipophilicity as the drug must go across the endothelium of the vessels and be eliminated by uveal blood flow (Urtti 2006) (Fig. 3.1).

References

Ahmed I (2003) The noncorneal route in ocular drug delivery. In: Mitra A (ed) Ophthalmic drug delivery systems. Marcel Dekker, New York

Ahmed I, Patton TF (1985) Importance of the noncorneal absorption route in topical ophthalmic drug delivery. Investig ophthalmol Vis Sci 26(4):584–587. Available at http://www.ncbi.nlm.nih.gov/pubmed/3884542. Accessed 20 March 2015

Anon (2013) Principles of metabolism in health and diseases. Springer Science & Business Media. Available at http://books.google.com/books?hl=en&lr=&id=75_IAwAAQBAJ&pgis=1. Accessed 21 March 2015

Anon (2008) Surgery for the dry eye: scientific evidence and guidelines for the clinical management of dry eye associated ocular surface disease. Karger Medical and Scientific Publishers. Available at http://books.google.com/books?hl=en&lr=&id=TzoAgrAjXYwC&pgis=1. Accessed 1 April 2015

Barar J et al (2009). Ocular drug delivery; impact of in vitro cell culture models. J Ophthalmic Vis Res 4(4):238–252. Available at http://www.pubmedcentral.nih.gov/articlerender.fcgi?artid=3498862&tool=pmcentrez&rendertype=abstract. Accessed 20 March 2015

Benitez-del-Castillo JM, Lemp MA (2012) Ocular surface disorders, JP Medical Ltd. Available at http://books.google.com/books?hl=en&lr=&id=7NCQXp2KAd0C&pgis=1. Accessed 7 April 2015

Bron A, Tiffany J, Gffney E (2008) Compartmental factors influencing tear film osmolarity. Acta Ophthalmol 86. Available at http://doi.wiley.com/10.1111/j.1755-3768.2008.4333.x. Accessed 1 April 2015

Cunha-Vaz, JG (1997) The blood-ocular barriers: past, present, and future. Documenta ophthalmologica. Adv Ophthalmol 93(1–2):149–157. Available at http://www.ncbi.nlm.nih.gov/pubmed/9476613. Accessed 20 March 2015

Cunha-Vaz JG (1976) The blood-retinal barriers. Doc Ophthalmol 41(2):287–327. Available at http://link.springer.com/10.1007/BF00146764. Accessed 2 April 2015

Dartt D, Hodges R, Zoukhri D (2006) Tear and their secretion. In: FJ (ed) The biology of the eye. Academic Press, New York, pp 21–82

Duvvuri S, Majumdar S, Mitra AK (2003) Drug delivery to the retina: challenges and opportunities. Exp Opin Biol Ther 3(1):45–56. Available at http://informahealthcare.com/doi/abs/10.1517/14712598.3.1.45. Accessed 20 March 2015

Egger SF et al (1994) Bacterial contamination during extracapsular cataract extraction. Ophthalmologica 208(2):77–81. Available at http://www.karger.com/Article/FullText/310457 . Accessed 21 March 2015

Freddo T (2001) Shifting the paradigm of the blood-aqueous barrier. Ex Eye Res 73:581–592

Gardner TW et al (1999) The molecular structure and function of the inner blood–retinal barrier. Doc Ophthalmol 97:229–237

Gipson IK (2004) Distribution of mucins at the ocular surface. Exp Eye Res 78(3):379–388. Available at http://www.sciencedirect.com/science/article/pii/S0014483503002045. Accessed 1 April 2015

Grandolfo M, Rindi A, Sliney DH (eds) (1991) Light, lasers, and synchrotron radiation. Springer US, Boston, MA. Available at http://link.springer.com/10.1007/978-1-4899-0661-8. Accessed 21 March 2015

Hämäläinen KM et al (1997) Characterization of paracellular and aqueous penetration routes in cornea, conjunctiva, and sclera. Investig Ophthalmol Vis Sci 38(3):627–634. Available at http://www.ncbi.nlm.nih.gov/pubmed/9071216. Accessed 20 March 2015

Hornof M, Toropainen E, Urtti A (2005). Cell culture models of the ocular barriers. Eur J Pharm Biopharm (Official journal of Arbeitsgemeinschaft für Pharmazeutische Verfahrenstechnik e.V) 60(2):207–225. Available at http://www.sciencedirect.com/science/article/pii/S0939641105000214. Accessed 20 March 2015

Jaleh B, Yadollah O (2008). Bioelectrical and permeability properties of brain microvasculature endothelial cells: effects of tight junction modulators. J Biol Sci 8:556–562. Available at http://scialert.net/abstract/?doi=jbs.2008.556.562

Janzer RC, Raff MC (1987) Astrocytes induce blood-brain barrier properties in endothelial cells. Nature 325(6101):253–257. Available at http://dx.doi.org/10.1038/325253a0. Accessed 2 April 2015

Macha S, Mitra A (2003) Overview of ocular drug delivery. In: Mitra A (ed) Ophthalmic drug delivery systems. Marcel Dekker, New York, pp 1–12

Maïssa C, Guillon M (2010) Tear film dynamics and lipid layer characteristics–effect of age and gender. Contact Lens Anterior Eye : J Br Contact Lens Assoc 33(4):176–182. Available at http://www.sciencedirect.com/science/article/pii/S1367048410000111. Accessed 1 April 2015

Marmor MF (1998) Structure-function and disease of the retinal pigment epithelium. In: Marmor MF, Wolfensberger TJ (eds) The retinal pigment epithelium. Oxford University Press, New York, pp 3–9

Mathers W (2004) Evaporation from the ocular surface. Exp Eye Res 78(3):389–394. Available at http://www.sciencedirect.com/science/article/pii/S0014483503001994. Accessed 1 April 2015

Maurice DM, Mishima S (1984) Ocular pharmacokinetics. In: Sears ML (ed) Handbook of experimental pharmacology. Springer, Berlin, pp 16–119

Mishima S, Maurice DM (1961) The oily layer of the tear film and evaporation from the corneal surface. Exp Eye Res, 1(1):39–45. Available at: http://www.sciencedirect.com/science/article/pii/S0014483561800067. Accessed 1 April 2015

Mitra A, Anand B, Duvvuri S (2006) Drug delivery to the eye. In: Fischbarg J (ed) The biology of the eye. Academic Press, New York, pp 307–351

Nichols KK et al (2011) The international workshop on Meibomian Gland dysfunction: executive summary. Investig Ophthalmol Vis Sci 52(4):1922–1929. Available at http://www.iovs.org/content/52/4/1922.short. Accessed 12 March 2015]

Pitkänen L et al (2003) Vitreous is a barrier in nonviral gene transfer by cationic lipids and polymers. Pharm Res 20(4):576–583. Available at http://link.springer.com/article/10.1023/A%3A1023238530504. Accessed 2 April 2015

Purves D et al (2001) Anatomy of the eye. Available at: http://www.ncbi.nlm.nih.gov/books/NBK11120/. Accessed 21 March 21 2015

Quinn RH, Miller SS (1992) Ion transport mechanisms in native human retinal pigment epithelium. Investig ophthal Vis Sci 33(13):3513–3527. Available at http://www.ncbi.nlm.nih.gov/pubmed/1334477. Accessed 20 March 2015

Rajasakul Y, Robinson JR (1989) Transport mechanisms of the Cornea: characterization of barrier perselectivity. Int J Pharm 55:237–246

Rolando M, Zierhut M (2001) The ocular surface and tear film and their dysfunction in dry eye disease. Surv Ophthalmol 45:S203–S210. Available at http://www.sciencedirect.com/science/article/pii/S0039625700002034. Accessed 12 March 2015

Schoenwald RD, Huang HS (1983) Corneal penetratiion behavior of beta-blocking agents I: physicochemical factors. J Pharm Sci 72(11):1266–72

Selvin B (1983) Systemic effects of topical ophthalmic medications. South Med J 76:349–358

Stewart PA, Tuor UI (1994). Blood-eye barriers in the rat: correlation of ultrastructure with function. J Comp Neurol 340(4):566–576. Available at http://www.ncbi.nlm.nih.gov/pubmed/8006217. Accessed 2 April 2015

Sunkara G, Kompella U (2003) Membrane transport processes in the eye. Ophthalmic drug delivery systems. Marcel Dekker, New York, pp 13–58

Urtti A. (2006) Challenges and obstacles of ocular pharmacokinetics and drug delivery. Adv Drug Del Rev 58(11):1131–1135. Available at http://www.sciencedirect.com/science/article/pii/S0169409X06001578. Accessed 6 January 2015

Urtti BA, Salminen L (1993) Minimizing systemic absorption of topically administered ophthalmic drugs. Surv Ophthalmol 37:435–456

Winter KN, Anderson DM, Braun RJ (2010) A model for wetting and evaporation of a post-blink precorneal tear film. Math Med Biol : J IMA 27(3):211–225. Available at: http://imammb.oxfordjournals.org/content/early/2009/10/27/imammb.dqp019.short. Accessed 1 April 2015

Yokoi N, Bron AJ, Georgiev GA (2014) The precorneal tear film as a fluid shell: the effect of blinking and saccades on tear film distribution and dynamics. Ocul Surf 12(4):252–266. Available at http://www.sciencedirect.com/science/article/pii/S1542012414001220. Accessed 7 April 2015

Chapter 4
Drug Delivery to Specific Compartments of the Eye

Lunawati Bennett

Abstract Due to different barriers that eye possesses, its complicated anatomical structure, and its small absorptive surface, drug delivery to specific parts of anterior segment or posterior segment have difficult challenges to overcome. Barriers exist at different compartment of the eye, drug delivery to different parts, and available ophthalmic formulations and those that are still in research will be discussed. This chapter will discuss drug delivery to specific parts of the eye like the anterior and posterior portion, vitreous humor, retina, choroid and others.

Keywords Compartment of the eye · Posterior segments · Anterior segments · Drug delivery · Ocular

Introduction

The eye is a unique organ. Drug delivery to various sites to treat eye-related diseases has been a difficult challenge to overcome. Despite several ocular formulations available on the market, no single route of administration or formulation has gained wide acceptance in treating ocular diseases in the anterior or the posterior segment of the eye.

Several known barriers of anterior and posterior segments include: (a) the "inner and outer BRB" (blood-retinal barrier) separate the retina and vitreous from the systemic circulation; (b) the inner limiting membrane that controls the exchange and entry of drug particles from the vitreous to the retina; (c) the BAB (blood-aqueous barrier) that limits the transport of drug from the blood to the inner part of the eye; (d) intact structure of corneal epithelium with desmosomes and tight junction that resist passage of most ophthalmic drugs due to layers of hydrophobic epithelium, hydrophilic stroma, and hydrophobic endothelium; and (e) the tear film

L. Bennett (✉)
School of Pharmacy, Union University, 1050 Union University Drive,
Jackson, TN 38305, USA
e-mail: llbennett@uu.edu

R.T. Addo (ed.), *Ocular Drug Delivery: Advances, Challenges and Applications*,
DOI 10.1007/978-3-319-47691-9_4

that forms a muco-aqueous barrier which continuously wash away drugs from the anterior surface of eye (Cholkar et al. 2013; Rathore and Nema 2009).

Anterior Segment Delivery

The anterior segment which occupied one-third of the eye is responsible for collecting and focusing light is divided into the cornea, crystalline lens, conjunctiva, iris, ciliary body, tear film and fluid filled aqueous humor. There are several challenges to deliver ophthalmic drugs to the anterior segment and these include: (1) static barriers (corneal epithelium, corneal stroma, BAB); (2) dynamic barriers (conjunctival blood flow, lymph flow, tear drainage); (3) metabolic barriers due to the presence of enzyme esterase; and (4) barriers due to efflux proteins pumps such as P-glycoprotein (P-gp), multi-drug resistance protein (MDRP), and breast cancer resistance protein (BCRP) on the eye membrane (Cholkar et al. 2013; Yavuz et al. 2013).

Common diseases of anterior segment include open-angle glaucoma (OAG), closure angle glaucoma (CAG), cataract, and dry eyes. Many topical formulations are capable of delivering drugs to the anterior segment of the eye for treatment of these diseases, but due to the barriers the drug encounters and the small amount of the drug being delivered, the drug concentration at the intended site is usually minimal (Baranowski et al. 2014). Therefore, designing a new drug delivery system that can efficiently target the anterior tissues, generate high drug levels, and maintain prolonged and effective concentration with no or minimal side effects is the major focus of current research in ophthalmic formulations.

To extend bioavailability of ophthalmic drugs with contact time to the cornea, some viscosity-enhancing polymers such as polyvinyl alcohol (PVA), poloxamers, cellulose, hyaluronic acid, carbomers (weakly cross-linked acrylic acids), polysaccharide, gum and xanthan gum are usually added (Gaudana et al. 2009). The viscosity of the drug can be made in the range of 15 to 150 mPas to increase the drug penetration to the cornea and to resist lacrimal drainage when residing in the lower conjunctiva cul-de-sac. Hydrophilic polymers exhibit mucoadhesive properties by creating a non-covalent bond with mucin in the conjunctiva, therefore causing longer contact with eye surface. The polymers in this class are polyacrylic acid (PAA), carboxyl methyl cellulose (CMC), lectins, and chitosan. The combination of PAA and CMC blends provides excellent and longer mucoadhesion effect to the eye membrane and increases the drugs' biocompatibility profiles (Sechoy et al. 2000). Penetration enhancing substances that work as chelating agents such as benzalkonium chloride, surfactants, and bile acid salts are sometimes added to the ophthalmic drugs to modify corneal epithelium absorption. However, these excipients can also cause toxicity; therefore, the lowest concentration should be used which also limit their use in ophthalmic preparations.

Another important useful chemical to be added to ophthalmic drugs is cyclodextrins (CDs), cyclic oligosaccharides that form a complex with the active ingredients of an ophthalmic drug by increasing the water solubility of hydrophobic drugs without changing their structures. As carriers, about less than 15 % of 2-hydroxy propyl-β-cyclodextrin (HPβ-CD) can be added to keep hydrophobic drugs in solution and to transport the drug into the cornea without causing irritation to the eye. CD, HPβ-CD and various forms of CDs have been added to highly hydrophobic molecules such as non-steroidal anti-inflammatory drugs (NSAIDs), corticosteroids to make these active ingredients be more hydrophilic in ophthalmic formulations (Baranowski et al. 2014; Gaudana et al. 2009; Loftsson and Stefansson 2002).

Delivery to Cornea

Ophthalmic drugs are usually delivered through the cornea, conjunctiva or sclera pathways. The corneal pathway involves entry of drugs into the cornea followed by the drug passing through interior tissues such as iris, aqueous humor, lens, and iris-ciliary body (Kompella et al. 2013; Gong et al. 2009). The cornea is a multi-layered, transparent, avascular, highly innervated sensitive tissue that is in close proximity to conjunctiva (Cholkar et al. 2013). Figure 4.1 shows position of the cornea, conjunctiva, tear film, corneal epithelium and aqueous humor of anterior segment.

The cornea is important for maintaining eye tissue integrity and function. It is composed of five different layers: the outer lipophilic epithelium, the Bowman's

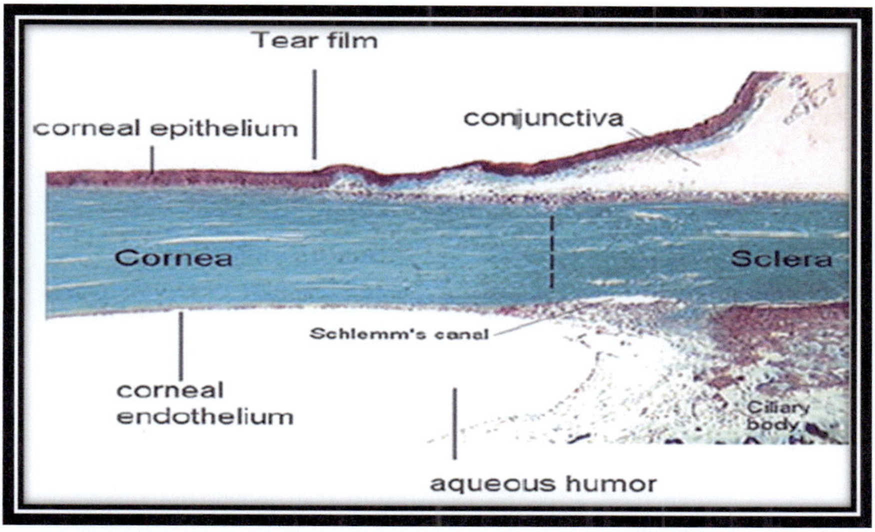

Fig. 4.1 Anterior ocular tissue and fluid. Available online at www.libertpub.com/jop. With permission from Cholkar et al. (2013)

layer, the hydrophilic stroma, the Descemet's membrane and the endothelium. Epithelial cells are highly lipophilic with a tight junction that prevents small hydrophilic drugs entering into ocular tissue, while hydrophobic drugs can cross the lipophilic epithelium but the drug can't penetrate further into deeper tissue due to the presence of hydrophilic stroma. Another barrier affecting drug delivery is the presence of transmembrane efflux pumps such as P-gp, MDRP, and BCRP on the cornea surface which cause ophthalmic drug absorption to be minimal (Cholkar et al. 2013; Gaudana et al. 2009). At physiological pH, the cornea is negatively charged. To prevent drug resistant, positively charged ophthalmic are better formulations for delivering a drug to the cornea (Gaudana et al. 2012). Drugs can also be lost via the conjunctiva, episcleral blood, and lymphatic flow (Hosoya et al. 2005).

Certain strategies can be employed to improve the bioavailability of the drug by formulating the drug into a prodrug, adding enhancers, additives or thermo-sensitive gel, using contact lenses, and delivering with Nanosystems, to name a few, so the drug can penetrate and stay longer in the cornea or precorneal area with fewer side effects.

Prodrugs

Prodrugs are bio reversible derivatives of drug molecules and are designed to be therapeutically inactive until enzymatic and/or chemical bio reversion occurs. These derivatives are generally synthesized by chemical conjugation of a specific moiety to the parent drug via an ester, amide, or other enzymatically cleavable linkages. Prodrug containing ester or amide linkages can undergo varying esterase and amide-mediated hydrolysis, which are present in the iris-ciliary body, aqueous humor, and cornea, so the drug can permeate through the cornea/conjunctiva and enter into the aqueous humor, iris and ciliary body (Kaur and Kanwar 2002; Majumdar and Mitra 2006).

Prodrugs increase drug permeability, selectivity and site specificity. Several prodrug derivatives available on the markets include analogs of prostaglandin (PGF2α) such as latanoprost, travoprost, isopropyl unoprostone, ethyl amide bimatoprost which have improved by 4–83 folds in their corneal permeation (Kompella et al. 2010). These drugs are used for the treatment of OAG or CAG. By changing to a prodrug, timolol a lipophilic drug which usually exhibits a high incidence of cardiovascular and respiratory side effect has been shown to increase the corneal permeability by 2–3 folds, increase drug concentration in aqueous humor by 4–6 folds, with a reduction for up to 2 folds in its side effects. Epinephrine, phenylephrine, and pilocarpine have also been formulated as prodrugs. For example dipivefrine, a prodrug of a diester of pivalic acid and epinephrine was shown to be 17-fold higher in their permeability to the cornea than the parent drug epinephrine. Dipivefrine at 0.1 % is comparable to 2 % epinephrine eye drops used to decrease intraocular pressure (IOP) for the treatment of glaucoma with less adverse effects (Jarvinen and Jarvinen 1996).

Other prodrugs that have been successfully used to deliver drugs to anterior segment include ganciclovir, dexamethasone, and flurbiprofen used as antiviral, anti-inflammatory with improvement in their efficacy, corneal permeability, and less side effects (Rautio et al. 2008; Hellberg et al. 2003).

Additive, Enhancers and Thermo-Sensitive Gel

Viscosity enhancing polymer can be incorporated into the ophthalmic formulations to improve residence time in the precorneal area and to increase absorption across the cornea. These polymers that have been successfully used are hydroxyl propyl methyl cellulose (HPMC), hyaluronic acid, polyvinyl alcohol (PVA), hydroxyl ethyl cellulose (HEC) and methylcellulose.

Another approach is embedding drug into a thermo-sensitive gel. For example, a dexamethasone-loaded poly (lactide-co-glycolide)-polyethylene glycol-poly (lactide-co-glycolide) (PLGA-PEG-PLGA) was found to be 7 folds higher in concentration and stayed longer in the precorneal than its aqueous eye drops formulation. Many other in situ polymeric gelling agents such as chitosan, HPMC, poloxamer, poly (hydroxyethyl methacrylate) p(HEMA), and others have been explored for the topical ocular application (Baranowski et al. 2014; Rajasekaran et al. 2010). These polymeric gelling agents can also be used in combination with nano formulations (*see below*). See also Chap. 5 under "*semi-solid ophthalmic forms*".

Contact Lenses

Contact lenses are hard or soft polymeric devices designed to fit directly onto the cornea to correct refractive abnormalities. Hydrogel contact lenses with nanoparticles can hold a large volume of aqueous solutions. For example, adding polymeric gelling agent to norfloxacin, or gentamicin which is used to treat microbial infection cause these drugs to stay between 8 h to up and 3 weeks longer than conventional eye drops (Gaudana et al. 2012; Chauhan and Gulsen 2012).

Nano Formulations (NFs) Delivery

Cyclosporine (CsA) is a highly lipophilic drug with poor aqueous solubility. It has been used to treat dry eye (keratoconjunctivitis) and allergic inflammation by inhibiting inflammatory cytokine production and by inhibiting eosinophils and mast cells activation and release. CsA has also been tried to treat other conditions such as anterior disease uveitis and atopic keratoconjunctivitis and posterior disease blepharitis; however, CsA induces a burning sensation and redness after topical instillation. Cationic, anionic, and oily CsA emulsion were studied to reduce inflammation and stinging sensation and to improve its safety and tolerability

(Ako-Adounvo 2014; Lallemand et al. 2012). Restasis® (0.05 % CsA nanoemulsion, Allergan) is commercially FDA approved for dry eye.

Several other NFs have been researched thoroughly such as: (1) nanoparticles incorporated ibuprofen, indomethacin, CsA, rapamycin, prednisolone, or gatifloxacin with PLGA, chitosan, or hyaluronic acid; (2) nanosuspension incorporated indomethacin, diclofenac, rofecoxib with sesame oil, pluronic F68, HPMC; (3) nanoemulsion incorporated CsA, indomethacin into chitosan, kelcogel; (4) nanomicelles incorporated dexamethasone, pilocarpine, cyclosporine, rapamycin with pluronic, chitosan, vitamin E; (5) cubosomes incorporated dexamethasone; and (6) liposomes incorporated dexamethasone into polyethylene glycol. See also Chap. 5 for different NFs formulations (Rathore and Nema 2009; Gaudana et al. 2009). With NFs delivery, drugs were shown to stay at the precorneal or corneal longer and at higher concentrations than aqueous eye drops.

Delivery to Conjunctiva

Conjunctiva pathway involves drug permeability across conjunctiva followed by entry into the sclera, choroid, retinal pigment epithelium, and retina. There are several corneal diseases caused by inflammation, viral, bacterial or fungal infections, degenerative diseases, neovascularization, and traumatic conditions. The conjunctiva is the first layer of anterior segment providing barriers to ophthalmic drugs due to: its tissue composition, the presence of enzymes esterase that break down drugs, the efflux pumps that cause drug efflux from the cell cytoplasm, high blood flow, and the presence of tear film. The conjunctiva is a thin, transparent, elastic, highly vascularized tissue lining the upper and lower eyelid and sclera, covering almost 80 % of the ocular surface. It consists of two layers: the outer epithelial layer which is about 2–10 layers depending on its anatomical location and the inner stroma layer. The outer epithelial has a tight junction that impedes passive diffusion of drugs across the cell layers. The inner stroma is richly supplied with nerves, blood, and lymph vessels and its contact with inner sclera can cause barrier to hydrophobic drugs delivery (Cholkar et al. 2013; Hosoya et al. 2005).

Conjunctiva lymph and blood circulation can act as a dynamic barrier for absorption of drugs. Another significant impediment for drug absorption is the tear film turnover. Conjunctiva secrets goblet cells responsible for mucin and tear formation. Tear film protects the corneal epithelial layer by preventing the eye from dehydration. Tear film thickness is about 3–9 μm and the pre-corneal pocket also called cul-de-sac can hold tear volume of about 7 μL. The tear film is composed of 3 layers: an outer lipid layer, a middle aqueous layer, and inner mucin layer. Tear fluid secretes electrolytes, glucose, immunoglobulin, lysozymes and lactoferrin which aid in lubrication, nourishment, maintenance and repair of corneal epithelium. Only about 10–20 % of the drug is available for absorption, the other 80–90 % are lost in the nasolacrimal duct (Cholkar et al. 2013; Sechoy et al. 2000).

Compared to corneal drug delivery, research of conjunctival drug delivery is relatively unexplored. The lack of research using conjunctival as a delivery system is somewhat surprising, given that the conjunctiva is always involved whenever topical application are used to treat anterior segment diseases. The prodrug approach to conjunctival drug delivery has not been exploited to a great extent relative to the drug delivery to the cornea. Several methods of delivering drugs to conjunctiva listed below and in Chap. 5.

NFs Delivery

NFs drug delivery to conjunctiva has been successfully done in in vitro models such as rabbit, bovine, porcine. After passing the conjunctiva, the NFs can reach the posterior segment via diffusion through the sclera, vitreous or through blood circulation to the choroid due to the size of NFs which are usually smaller than normal drugs. See Chap. 5 for more on NFs.

New Ophthalmic Delivery System (NODS)

Recently, NODS was patented by Smith and Nephew Pharmaceuticals Ltd. A film containing the drug was introduced into the conjunctival sac and the drug dissolved in the tear fluid, releasing the active ingredient (Rajasekaran et al. 2010).

Mini Tablets

Mini tablets are biodegradable, solid drug forms, that after application to conjunctival sac, transit into gels, which extends the time period of contact between active ingredients and the eyeball surface, which increase the active ingredients bioavailability. The mini tablets usually have polymers such as HPMC, HEC, ethyl cellulose, magnesium stearate, carbomer, and others. Mini tablets are developed by applying direct compression or an indirect method using dry granulation stage that can increase flow properties of the powder. Active ingredients with mini tablets that have been developed include using piroxicam, timolol, ciprofloxacin, gentamicin and acyclovir for treatment of corneal inflammation and infections (Rathore and Nema 2009; Gaudan et al. 2009) (Fig. 4.2).

In conclusion, topical delivery system to anterior segment:

- *Barriers involved in topical delivery*: cornea, inner and outer BRB, blood dilution of the drug.
- *Elimination of drugs*: tear wash out, nasolacrimal drainage, choroid, and conjunctival blood flow.
- *Advantages of these formulations*: high patient compliance, less systemic adverse effects.

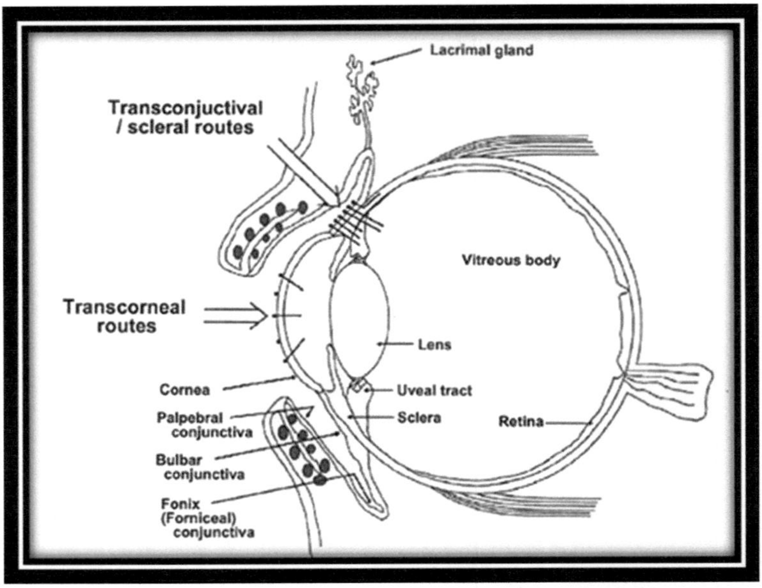

Fig. 4.2 Cross-section of eye and various drug absorption routes. With permission from Hosoya et al. 2005

- *Disadvantages of these formulations*: low bioavailability, high dose requirement, systemic adverse drug reactions, small retention time, blur vision, drug losses, pre-corneal drug losses, drainage through the nasolacrimal duct, irritation, and low bioavailability.

Posterior Segment Delivery

Posterior segment occupied two-third of the eye responsible for detecting light. The segment includes sclera, choroid vessels, Bruch's membrane, retinal pigmented epithelium (RPE), retina, macula, optic nerve, and fluid-filled vitreous humor. The barrier to posterior delivery includes: retina, vitreous body and humor, BRB, sclera, and choroid. Penetration of drugs to posterior segment such as retina is usually negligible due to various pre-corneal/cul de sac or anatomical factors such as solution drainage, blinking, tear film, tear turnover, lipophilic nature of corneal epithelium, hydrophilic nature of corneal stroma, and the presence of tight junction (Kompella et al. 2013; Pescina et al. 2011).

Of the total debilitating diseases, 55 % of the diseases affect the posterior segment with only 5 % ophthalmic drugs available for posterior treatment. Common diseases affecting posterior segments of eye include: age-related macular degeneration (AMD), proliferative vitreoretinopathy (PVR), uveitis, diabetic retinopathy, diabetic macular edema, viral retinitis, posterior uveitis, choroid neovascularization (CNV), retinitis pigmentosa, cytomegalovirus (CMV) retinitis, and other ocular neovascular disease (Donovan 2004; Morrison and Khutoryanskiy 2014).

Sclera occupies a large total surface area of the eye (about 95 %), is hypocellular and more permeable than cornea (Yasin et al. 2014). It is highly permeable to small or high molecular weight ophthalmic drugs. The permeability of drug across sclera is inversely proportional to the molecular weight and lipophilicity of the drug. Large proteins like albumin or large macromolecules like dextran, when injected into the suprachoroidal space, were able to diffuse across the sclera and can be detected in the extraocular tissue. The sclera is a relatively hydrophilic tissue, made of collagen fibrils matrix with fluid-filled channels that is permeable to a wide range of solutes, and glycosaminoglycan that are negatively charged and can bind positive-charged drugs. Permeability across sclera was greater with anionic than cationic ophthalmic drugs (Pescina et al. 2011).

Retina is rich with glycosaminoglycan layer that separates the underlying pigmented retinal layer from the vitreous humor. It is a multilayered structure with inner and outer limiting membranes enveloping inter-photoreceptors forming a significant barrier for drugs penetration (Fig. 4.3). The glycosaminoglycan layer binds to cationic molecules and limits their transport through the retina. The charge of the ophthalmic drugs plays more important limiting factor than molecular size of the drugs when transport across the neural retinal barrier (Kaur and Kakkar 2014).

The inner meshwork pores ranging in sizes preventing macromolecules from the vitreous to the retina. The vitreous body is mainly composed of water 98 %, very small amount of solid components such as collagen and glycosaminoglycan. Glycosaminoglycan impedes free diffusion of high molecular weight ophthalmic drugs through the vitreous.

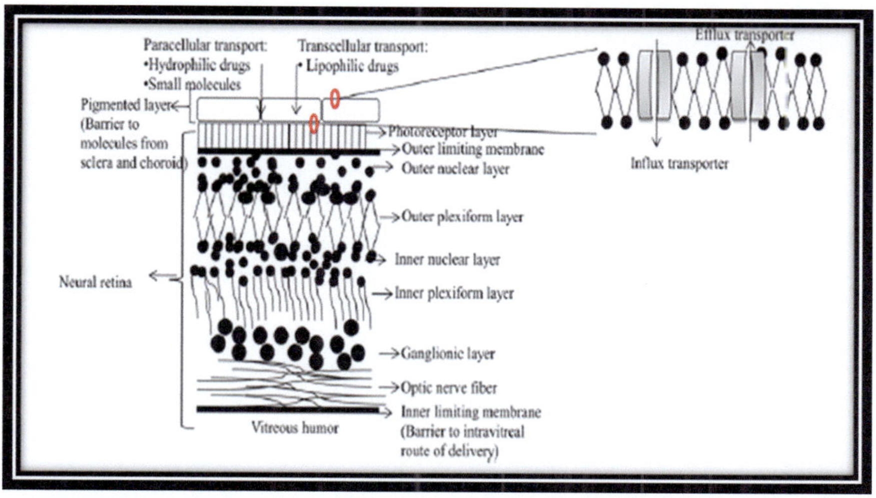

Fig. 4.3 Schematic representation of different layers of retina and transport across the RPE. With permission from Kaur and Kakkar (2014)

BRB which composed of tight junction of retinal endothelial blood vessels and RPE also act as a significant barrier for drug absorption into the retina and vitreous. RPE restricts ophthalmic drugs after systemic administration, limit drug absorption from the choroid to the retina, limit hydrophilic drugs, and macromolecules permeability. Additionally, small lipophilic drugs and/or cationic molecules can bind to melanin in the RPE which further decrease the penetration of the drugs. The choroid and RPE are relatively lipophilic. The permeability across the RPE in general increases with increase drug lipophilicity until the plateau is reached (Cholkar et al. 2013; Kaur and Kakkar 2014). It is more complicated to deliver drugs to retina because of the choroid and RPE (Yavuz et al. 2013).

Vitreous extend from the lens to the retina consists of a highly hydrated gel-like extracellular matrix. Controlled release drug delivery using an injection into the vitreous (intravitreal route) offer drug delivery to the target site with minimal surgical invasion. Vitreous has also been used for drug delivery using biodegradable or non-biodegradable implants.

Choroid especially Bruch's membrane are highly vascularized part of the eye, but also provides a significant barrier to macromolecule drugs. Choroid-Bruch's membrane is a more effective barrier for lipophilic than hydrophilic drugs. Transport molecule across sclera-choroid—RPE was the least for highly lipophilic when compared to hydrophilic molecules (Kompella et al. 2013; Gong et al. 2009). Pigmentation is present in the choroid and RPE. Pigmented nature of the choroid and its affinity for lipophilic solutes can also limit the drug to reach the retina. People with eumelanin possess black or brownish color eye, while pheomelanin causes reddish and yellowish eye. Binding of drugs to pheomelanin or eumelanin differ significantly depending upon the drug characteristics. Pigments are polyanionic with several carboxyl groups and semiquinones that have a tendency to bind lipophilic and cationic drugs strongly through hydrophobic and electrostatic interactions; therefore, reducing transport of the lipophilic drug across the sclera-choroid.

Other barriers to effective treatment of posterior segment include the presence of P-gp, MRP, other efflux transporters found in the membrane of RPE which efflux molecules out of the cytoplasm and limit permeability of drugs from choroid to retina and vitreous chamber (Ito and Walter 2013; Wallace et al. 2014).

Various routes administration for posterior segment delivery includes: topical, systemic, periocular, and intraocular routes. Topical drug deliveries have shown poor clinical success for treatment of posterior disease due to an only small fraction of drugs reaching the intended area. Targeting drug by the systemic method such as by intravenous is still a limited method due to undesirable systemic side effects and the presence of BRB. After systemic administration, drugs can penetrate and diffuse through any leaky vessels into the choroid, but only small hydrophilic drugs can penetrate into the posterior segments. Intraocular and periocular drug delivery to the posterior segments is two commonly used methods to deliver drugs to posterior segments.

Intraocular drug delivery technique is intended to deposit the drug in the eye directly at the site of action. This method delivers drugs directly to the vitreous chamber. Several intravitreal deliveries such as: botulinum neurotoxin; intravitreal sustained release (SR) dosage forms having encapsulated botulinum toxin in the PLGA polymer; pulsatile release dosage form; water-in-oil emulsion and controlled-release microsphere-based formulation; were developed to extend the neurotoxin in the eye. Several methods of intraocular deliveries include: intrastromal, intracameral, suprachoroidal, subretinal and intravitreal. For detailed information see Chap. 5.

The advantage of the intraocular delivery is: shorten the distance drugs need to diffuse therefore increasing local concentration, reduce drug delivery to other off-target sites to decrease undesirable side effects, and bypass ocular epithelial and other ocular barriers to increasing bioavailability. The disadvantages are invasive methods, but repetitive injections can cause complication such as vitreous hemorrhage, retinal detachment, cataract, endophthalmitis, and inflammation (Donovan 2004).

Periocular drug delivery includes subconjunctival, sub-tenon, retrobulbar, peribulbar and posterior juxta scleral. The drugs given by these routes can be delivered to different layers of the eye. Periocular routes can potentially provide safer alternatives for delivering drugs to the retina and avoiding the risk of intraocular damage posed by intravitreal injection. By injecting the drugs outside the globe of the eye and in the proximity of sclera to increase better scleral permeability, the drug is delivered to the retina by diffusion through the sclera, choroid, and the RPE. To deliver to the retina, the molecule size of the drugs should be greater than 2000 nm. Drugs with molecule size of less than 200 nm will most likely retain at the periocular site of administration for prolonged periods. Although there is still increased the risk of systemic side effects, but it is considerably less than systemic delivery. This method might be good for delivery of long duration drug to treat posterior diseases since the drug can be delivered to the desired site in large concentrations while avoiding systemic side effects (Kaur and Kakkar 2014). Figure 4.4 shows different routes of drug delivery to the posterior segment. Figure 4.5 shows where periocular delivery is.

In conclusion, periocular and intraocular methods commonly used for posterior segment delivery.

Periocular Delivery

This delivery potentially provide safe alternative to the intravitreal injection, by delivering drugs to the retina (Kaur and Kakkar 2014);

- *Barriers involved in periocular delivery*: sclera, BRB, choroid.
- *Elimination of drugs*: conjunctival, choroidal blood, lymphatic flow, and periocular space.
- *Advantages of periocular delivery*: less painful than intravitreal method, most efficient route, achieve high therapeutic drug levels.

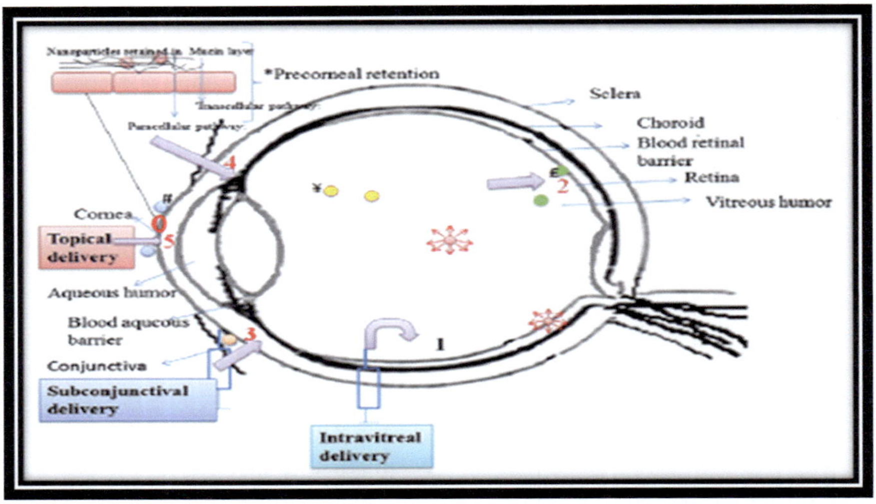

Fig. 4.4 Systemic proposed pathways to posterior segment by administration of Nanocarrier systems: *1* through intravitreal injection; *2* via BRB after systemic delivery; *3* trans-scleral route after subconjunctival or periocular delivery; *4* lateral non-corneal diffusion pathway after topical administration; *5* corneal diffusion pathway after topical administration. With permission from Kaur and Kakkar (2014). Slow drug release from the nano-structured carrier system acting as a depot in the vitreous and retinal cells. *, #, ¥, £ indicate uptake mechanisms of nano-carrier systems upon topical application. * is nanocarriers retained in the mucin layer and/or being mucoadhesive achieve increased pre-corneal time, # is adhesion to the corneal epithelium, ¥ is nano-carrier in the vitreous humor, £ is nano-carriers approaching retina

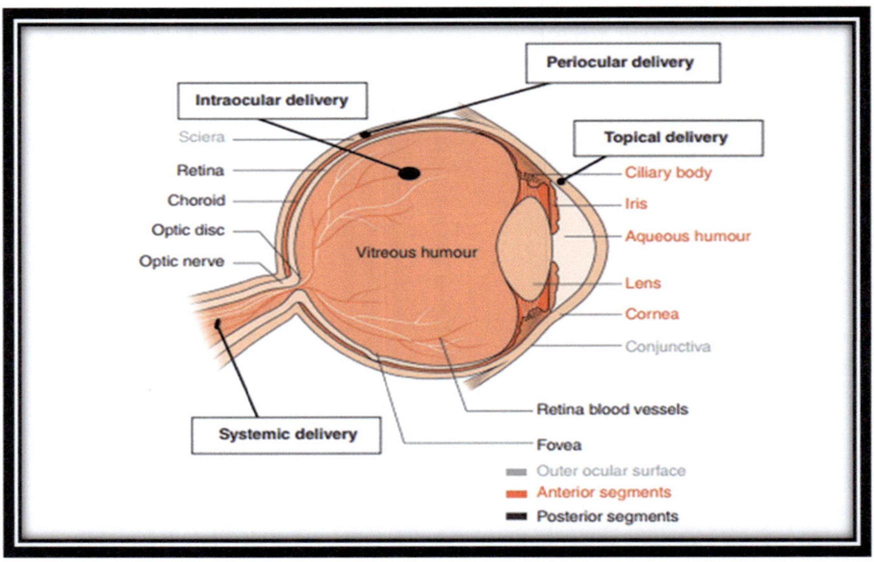

Fig. 4.5 Routes of administration for topical, intraocular, systemic and periocular. With permission from Shen et al. (2015)

- *Disadvantages of periocular delivery*: changes in tissue as a result of aging and pathophysiological alteration, a breakdown of BRB can significantly impact the delivery of drugs to the retina.

Intraocular Delivery

Intravitreal drug delivery is the most sought for the treatment of posterior eye diseases due to its proximity to the retina, choroid, and RPE.

- *Barriers involved in intravitreal delivery*: diffusion through the vitreous chamber, retina, and BRB.
- *Elimination of drugs*: movement to aqueous chamber and retina.
- *Advantages of intravitreal delivery*: local and direct delivery because the injection is in the proximity of the vitreous of the retina, providing highest bioavailability of the drugs.
- *Disadvantages of intravitreal delivery*: repeat injections, rapid elimination, serious adverse effects such as retinal detachment, hemorrhage, ocular hypertension and endophthalmitis, rapid clearance, systemic adverse effects, and tissue hemorrhage, poor patient compliance, painful procedure.

Future Innovative to Posterior Delivery Using Nano Formulation (NFS)

Because of the invasiveness of drug delivery to the posterior segment, it is important to design formulation that can maintain the therapeutic concentration in the posterior eye for prolonged periods, retention at the site of application; minimizing repetitive systemic, oral or parenteral injections; avoidance of irritation by improving solubility, stability, permeability of the drug; and are able to effectively deliver lipophilic or hydrophilic drugs. Generally, lipophilic drugs are more difficult to formulate as drops or injection for ocular delivery, while hydrophilic drugs show poor permeability across an ocular tissue. Nano formulation (NF) is designed to encapsulate hydrophobic or lipophilic drugs and carry the drugs as eye drops or injection across a series of barriers to the posterior eye segment effectively (Gupta et al. 2011; Wenger et al. 2011). Several forms of NFs have been mentioned in detailed in Chap. 5. Most common NFs to reach posterior segment of the eye are depicted in Fig. 4.6.

Conjunctival entry and transport of drugs can be enhanced using NFs. The drugs can stay in the choroid-RPE much higher and longer than drugs prepared in solution.

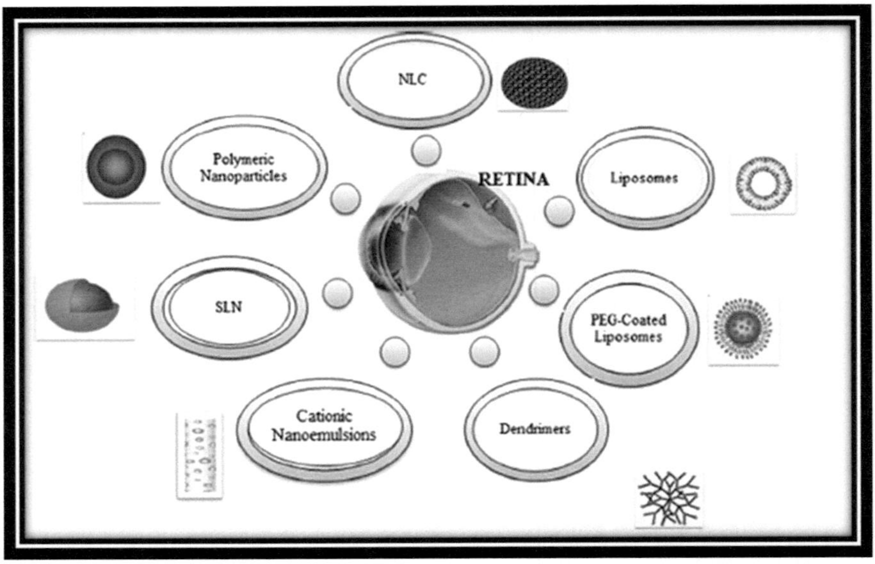

Fig. 4.6 Schematic illustration of the most common nanoparticles to reach the posterior segment of the eye. *PEG* polyethylene glycol; *SLN* solid lipid nanoparticles; *NLC* nanostructured lipid carriers. With permission Fangueiro et al. (2015)

IV administration of NF is also useful to target the disease site more specifically, enhance biodistribution to the target, and reduce requirements for using a high dose of active drugs (Gong et al. 2009; Kompella et al. 2013). Efficacy of NF of controlled release bevacizumab loaded with PEG and PLGA intravitreal injected for treatment of anti-angiogenic showed the drug stayed up to eight weeks longer than non- NF formulation (Ako-Adounvo et al. 2014; Wallace et al. 2014).

Additionally, NFs can be delivered using micro or nano electromechanical systems (M/NEMS) delivery. M/NEMs based fabrication use microneedles or implants with micro—and/or nanoparticles with sensor devices to monitor fluid flow. Micro-needles are micron sized syringe needles used as pain-free or reduce injection pain for delivering small or large molecular weight ophthalmic drugs. When micro-needles are used to puncture the globe for intraocular delivery, they by-pass ocular diffusional barriers and create transient transport pathways enhancing drug delivery within close proximity to the target tissue.

There is electrical component to monitor fluid within this system. Microneedles can be manufactured by laser drilling, polysilicon micro molding, or lithography electroforming replication. The microneedles can overcome barriers of the eye and deliver a drug to the sclera, the suprachoroidal space and beyond. The Minimal pressure of 250 kPa and a minimum microneedle length of 800 μm was required to penetrate the sclera for particle drug size 20–100 nm. For drug size 500–1000 nm, a minimum pressure of 300 kPa and a needle length of 1000 μm are required. Use of microneedle can overcome the scleral barrier and localize the drug to the posterior segment of the eye more efficiently (Gaudana et al. 2012; Ito and Walter 2013).

Summary

The incidence of chronic ocular diseases such as OAG, COG, AMD, cataract and other retinal degenerative disorders is expected to rise dramatically as people live longer with better health coverage. There is an urgency to develop a novel ocular delivery system that meets the needs of elderly people with visual impairment. While topical delivery may be ideal for treating anterior segment diseases, it may not be the most reliable drug delivery to the posterior segment. Several methods of posterior segment delivery usually involve more invasive drug delivery. There is a need for innovative drug delivery for different diseases of anterior and posterior.

References

Ako-Adounvo AM, Nagarwal RC, Oliveira L, et al (2014) Recent patents on ophthalmic nano formulations and therapeutic implications. Recent Pat Drug Deliv Formul 8(3):193–201

Baranowski P, Karolewicz B, Gajda M, Pluta J (2014) Ophthalmic drug dosage forms: characterization and research methods. The Sci World J. http://dx.doi.org/10.1155/2014/861904

Cholkar K, Patel SP, Vadlapudi AD, Mitra AK (2013) Novel strategies for anterior segment ocular drug delivery. J Ocu Pharm Ther 29(2):106–123

Chauhan A, Gulsen D (2012) Ophthalmic drug delivery system. US8273366

Donovan S (2004) Intravitreal botulinum toxin implant. US200475871

Fangueiro JF, Silva AM, Garcia ML, Souto EB (2015) Current nanotechnology approaches for the treatment and management of diabetic retinopathy. Eur J of Pharm Biopharm 95:307–322. doi:10.1016/j.ejpb.2014.12.023

Gaudana RJ, Gokulgandhi MR, Boddu SHS, Mitra AK (2012) Recent overview of ocular patents. Recent Pat Drug Deliv Formul 6(2):95–106

Gaudana R, Jwala J, Boddu SHS, Mitra AK (2009) Recent perspective in ocular drug delivery. Pharm Res 26(5):1197–1216

Gong X, Peng S, Wen W et al (2009) Design and fabrication of magnetically functionalized core/shell microspheres for smart drug delivery. Adv Funct Mater 19:292–297

Gupta H, Aqil M, Khar RK et al (2011) Biodegradable levofloxacin nanoparticles for sustained ocular drug delivery. J Drug Target 19(6):409–417

Hellberg MR, Ke TL, Haggard K et al (2003) The hydrolysis of the prostaglandin analog prodrug bimatoprost to 17-phenyl-trinor PGF2α by human and rabbit ocular tissue. J Ocul Pharmacol Ther 19:97–103

Hosoya K, Lee VHL, Kim KJ (2005) Roles of the conjunctiva in ocular drug delivery: a review of conjunctival transport mechanisms and their regulation. Eur J Pharm Biopharm 60:227–240

Ito YA, Walter MA (2013) Genetic and environmental mental contribution to anterior segment malformations and glaucoma. In: Rumelt S (eds) Glaucoma-basic and clinical aspects
Jarvinen T, Jarvinen K (1996) Prodrugs for improved ocular drug delivery. Adv Drug Del Rev 19 (2):203–224
Kaur IP, Kakkar S (2014) Nanotherapy for posterior eye diseases. J Control Release 193:100–112
Kaur IP, Kanwar M (2002) Ocular preparations: the formulation approach. Drug Dev Ind Pharm 28:473–493
Kompella UB, Amrite AC, Ravi RP, Durazo SA (2013) Nanomedicines for back of the eye drug delivery, gene delivery, and imaging. Prog Retin Eye Res 36:172–198
Kompella UB, Kadam RS, Lee VHL (2010) Recent advances in ophthalmic drug delivery. Ther Deliv. 1(3):435–456
Lallemand F, Daull P, Benita S, et al (2012) Successfully improving ocular drug delivery using the cationic nanoemulsion. Novasorb J Drug Deliv, 604204
Loftsson T, Stefansson E (2002) Cyclodextrins in eye drop formulations: enhanced topical delivery of corticosteroids to the eye. Acta Opthalmologica Scandinavica 80:144–150
Majumdar S, Mitra AK (2006) Chemical modification and formulation approach to elevated drug transport across cell membranes. Expert Opin Drug Delv 3:511–527
Morrison PWJ, Khutoryanskiy VV (2014) Advances in ophthalmic drug delivery. Ther Deliv 5 (12):1297–1315
Pescina S, Santi P, Ferrari G, Nicoli S (2011) Transscleral delivery of macromolecules. Ther Deliv 2:1331–1349
Rajasekaran A, Kumaran KSGA, Preetha JP, Karthika K (2010) A comparative review on conventional and advanced ocular drug delivery formulations. Int J Pharm Tech Res 2(1):668–674
Rathore KS, Nema RK (2009) An insight into ophthalmic drug delivery system. Int J Pharm Sci Drug Res 1(1):1–5
Rautio J, Kumpilainen H, Heimbach T (2008) Oridrugs: design and clinical applications. Nat Rev Drug Discov 7(3):255–270
Sechoy O, Tissie G, Sebastian C et al (2000) A new long-acting ophthalmic formulation of carteolol containing alginic acid. Int J Pharm 207:109–116
Shen HH, Chan EC, Lee JH et al (2015) Nanocarriers for treatment of ocular neovascularization in the back of the eye: new vehicles for ophthalmic drug delivery. Nanomedicine (Lond) 10:2093–2107
Wallace DM, Murphy-Ullrich JE, Downs JC, O'Brien CJ (2014) The role of matricellular proteins in glaucoma. Matrix Biolol 37:174–182
Wenger Y, Schneider RJ, Reddy GR et al (2011) Tissue distribution and pharmacokinetics of stable polyacrylamide nanoparticles following intravenous injection in the rat. Toxicol Appl Pharmacol 251:181–190
Yasin MN, Svirskis, Seyfoddin A, Rupenthal ID (2014) Implants for drug delivery to the posterior segment of the eye: a focus on stimuli-response and tunable release systems. J Control Rel 196:208–221
Yavuz B, Pehlivan SB, Unlu N (2013) Dendrimeric system and their applications in ocular drug delivery. The Sci World J. http://dx.doi.org/10.1155/2013/732340

Chapter 5
Topical Versus Systemic Ocular Drug Delivery

Lunawati Bennett

Abstract The eye is a very complex sensory organ consists of numerous structures function to coordinate sight properly. Several diseases related to eye include: (a) common inflammation and allergic reactions due to bacterial, viral, fungal or chemicals such as cytomegalovirus (CMV) retinitis and allergic rhinitis; (b) inflammatory and autoimmune disease such as scleritis and uveitis; (c) ocular neovascularization such as age-related macular degeneration (AMD), diabetic retinopathy (DR); and (d) retinal vein occlusion that can cause blindness if untreated or treated improperly. For years, ophthalmic formulations have been one of the most important, widely developed and challenging as pharmaceutical companies try to develop innovative drugs. Due to the complicated anatomical structure and a small absorptive surface of the eye, it is difficult to reach the eye compartment properly. Ophthalmologists still face challenges in treating different diseases of the anterior and posterior segments. Systemic, intraocular, and other methods of drug delivery are explained below with major emphasis on topical deliveries.

Keywords Topical delivery · Systemic delivery · Diseases of the eye · Intraocular delivery

Introduction

Systemic, extraocular (topical and subconjunctival), and intraocular (intrastromal, intracameral, subretinal, intravitreal), are the three main routes of delivering ophthalmic drugs. Each has its own advantages and disadvantages. Topical drug delivery is the most common route accounting for about 90 % of available formulations used for the treatment of conjunctivitis, uveitis, keratitis, scleritis, and others. The anatomical structure of the eye causes difficulty in delivering effective

L. Bennett (✉)
School of Pharmacy, Union University, 1050 Union University Drive,
Jackson, TN 38305, USA
e-mail: llbennett@uu.edu

R.T. Addo (ed.), *Ocular Drug Delivery: Advances, Challenges and Applications*,
DOI 10.1007/978-3-319-47691-9_5

drug concentration to the intended site and the ability of the drugs to stay longer to be effective. Some of these problems are: small absorptive surface of the eye, low transparency of the cornea, lipophilicity or hydrophilicity moieties of different epitheliums of the eye, low capacity of conjunctival sac that can only hold 30 μL of drugs, weak bonding of the drug with proteins contained in tear fluid, blinking which cause loss of the drugs, the presence of efflux transporter such as P-glycoprotein (P-gp) that efflux drugs out of the eye, and the presence of several barriers such as blood-aqueous barrier (BAB) and blood-retinal barrier (BRB) that prevent optimal drug delivery to the intended area of the eye (Gaudana et al. 2009; Jarvinen and Jarvinen 1996; Rajasekaran et al. 2010). Chapter 4 discussed in detail barriers facing drugs delivery to anterior or posterior segments of the eye.

Giving oral drugs using sustained-release (SR) formulation can be an advantage for patients' compliances in the treatment of lifelong disease such as glaucoma, AMD, but systemic undesirable adverse effects should also be expected (Morrison and Khutoryanskiy 2014; Martin 2012).

Intraocular delivery is a technique used to deposit drug to the eye directly using injection to the cornea (intrastromal), anterior segment (intracameral), or vitreous cavity in the center of the eye (intravitreal). These methods are effective in getting the ophthalmic drugs into the posterior segment of the eye such as for AMD treatment, but they are invasive procedures require sterile techniques with increased risks of retinal detachment, hemorrhage, endophthalmitis, and ocular hypertension (Bell et al. 2004; Jiang et al. 2009; Patel et al. 2011).

Systemic Delivery

Oral and parenteral injections are the two most common methods of delivery that can achieve systemic dosing. Oral delivery of large molecule drug is a non-invasive method to deliver a drug, but it has a limited penetration into the targeting eye tissue. The drug is also easily degraded in the gastrointestinal tract and become inactive after going through the first-pass metabolism in the liver; therefore, the amount of active drug that reaches the eye and its bioavailability is very minimal (less than 2 %). Small molecule drugs such as analgesics, antibiotics, and antiviral ophthalmics reach the intended site at low concentration (Kim et al. 2014). Formulating drug into SR such as for the treatment of glaucoma allow continuous and effective concentration of the drug at the intended site, but the systemic undesirable side effects are also greater and need to be considered.

Several systemic deliveries for the treatment of CMV retinitis, uveitis, and scleritis have been employed. CMV is an opportunistic virus that can infect ocular tissue such as the retina. This disease affects up to 25 % of patients with acquired immune deficiency syndrome (AIDS). Ganciclovir is the first line treatment that was initially given to the patient intravenously. Due to unwanted systemic side effects such as neutropenia, the systemic delivery has been replaced by intravitreal delivery (Yasin et al. 2014; Gaudana et al. 2010).

Uveitis is inflammation of the uvea that can cause blindness. It is divided into 4 types: (1) inflammation of the iris also called anterior uveitis, (2) inflammation of ciliary body also called intermediate uveitis, (3) inflammation of the choroid also called posterior uveitis, and (4) inflammation of all the part also called panuveitis. The current treatment is to suppress the inflammation by using topical or systemic immune suppressive such as methotrexate, cyclosporine A (CsA), cyclophosphamide, and biologics such as adalimumab or infliximab. Most of these drugs are given systemically. The drugs are less efficacious and have less specific effect at the intended site of the eye. It also causes systemic undesirable side effects which if severe enough prompt patients to discontinue taking the drug, thus causing a relapse of the disease (Yasin et al. 2014).

Scleritis is a serious inflammatory disease that can be due to infection, autoimmune or idiopathic. It can cause severe ocular pain, headache, periorbital pain, congestion of blood vessels and edema of the sclera, and occasionally blindness. Steroids are given systemically or topically have been used to treat this inflammation, but the systemic side effects cause steroids as non—preferable agents. Oral CsA 2–3.5 mg/kg/day have been used successfully in several case reports to treat uveitis and scleritis with minimal or no side effects. The ideal value of CsA ranges from 50 to 150 ng/ml to avoid systemic side effects and yet still achieving disease remission (Cholkar et al. 2015).

The choroid is the most common ocular site for breast or lung cancer to metathesis, due to its abundant vascular supply. Treatment option for choroidal metastases includes whole eye radiotherapy, systemic chemotherapy, immunotherapy, or hormone therapy. A large portion of choroidal metastases also expresses estrogen or progesterone receptors, making therapy with oral or parenteral injections as an effective option. For example, tamoxifen or aromatase inhibitors which are used to treat breast or lung cancer, also benefit to prevent cancer metathesis to choroid (Arepalli et al. 2015).

Following oral or parenteral injections, the BAB, BRB, and efflux transporter are obstacles that drugs encounter to enter the eye. Large size molecules are rarely used to deliver the drug to the eye due to poor absorption across many ocular barriers while increasing the dose of the drug can increase the severity of systemic side effects with the bioavailability of the drug is still very low. Targeting drugs to the posterior segment of the eye can be achieved by intravenous, but the drug penetration is limited due to BRB. After systemic administration, drugs can penetrate through any leaky vessels in the choroid and diffuse into the posterior segment. Numerous fenestrae are present in the endothelium of the choriocapillaris resulting in very little resistance for the transport of systemic solutes into the choroid. However, the presence of efflux transporters such as P-glycoprotein (P-gp) and multidrug resistance-associated proteins (MRP) found in the cell membrane of RPE can limit drug permeability from choroid to retina and vitreous chamber (Kompella et al. 2013).

Several attempts to improve drug delivery systematically using drug transporters have been done successfully in several animal models and ophthalmic cell lines. Cellular transporters play an important role in the disposition of drugs at the site of therapeutic action. An efflux transporter such as P-gp and MRP can limit the amount of drug reaching the target ocular tissue, thus limiting their efficacy, while an uptake or influx transporters present in ocular such as amino acid and peptide transporters are known as ASCT1A, peptide/histidine transporters (PHT1 and PHT2) can elevate the level of drugs in the target ocular tissue. Using influx transporter-targeted prodrugs, oral and systemically administered drugs can be distributed more efficiently. These molecules can interact with topical drugs by altering their local pharmacokinetics, safety, and efficacy. In the same way, topically administered drugs can affect the ocular distribution of orally or parenterally administered drugs by inhibiting the efflux transporters in the blood-ocular barriers, in which the oral/systemic drugs behaves as a substrate of transporters (Chemuturi and Yanez 2013). Further investigations are needed to modify the action of efflux and influx transporters system in the eye, so systemic ocular delivery, which usually have higher compliance and non-invasive route, can deliver drugs to the eye compartment with the highest bioavailability and lowest systemic side effects.

For systemic delivery to be successful, a relatively high drug concentration needs to be circulating in the blood plasma in order to achieve a therapeutically effective dose within the eye. Avoiding ocular barriers, altering influx systems, changing the formulations of oral or parenteral formulations are important innovations to make a systemic delivery to the eye efficient, with less systemic side effects, and minimal metabolism and clearance by the liver and kidney.

Topical Delivery

Topical delivery is the most accepted route accounting for approximately 90 % of aqueous ophthalmic formulations. Advantages of topical delivery are their relative simplicity to formulate, minimal storage of the drug, and ease of drug instillation thus increasing patients' compliances. Disadvantages include limited drug concentration at the eye if formulated as lipophilic drugs, significant loss at precorneal, the different barriers in the eye that cause low bioavailability at the intended site. There are several topical deliveries commercially available or in the research such as: (1) liquid forms (eye drops, ophthalmic solutions, micro or nanoemulsions); (2) semisolid forms (in situ gel, ointment); (3) solid forms (soft contact lenses, different inserts, mini tablets); and (4) other forms of delivery (sprays, iontophoresis, filter strips).

Liquid Ophthalmic Forms

Eye Drops

Eye drops are available as water and lipid solutions or suspensions. These forms are sterile, isotonic, with the pH equal to tear fluid (pH 7.4), and they may contain preservatives for multi-use packages. If the pH of the drugs is outside the range of 4–8, the drug can cause eye irritation and decrease in their bioavailability. The eye drops are easy to instill, but only a very small amount of the drugs are being absorbed into the target tissues. Therefore, it is important to apply large doses of drugs frequently to achieve the therapeutic doses, but large doses can cause more local and systemic side effects. (Baranowski et al. 2014; Gaudana et al. 2009; Jarvinen and Jarvinen 1996; Rajasekaran et al. 2010).

Ophthalmic Solutions

Ophthalmic solutions are sterile aqueous solution used to cleanse and rinse eye. It may contain excipients to regulate osmotic pressure, the pH, and viscosity of the products. The advantages are non—invasive, low systemic absorption, avoidance of first pass metabolism, ease of application with a small dose. Disadvantages include rapid drainage, rapid drug removal due to reflex blinking, and rapid degradation by enzymes such as esterases and amidases. Drugs should possess active molecules with a balance of hydrophilicity and hydrophobicity to generate longer time when in contact with cornea and conjunctiva. (Baranowski et al. 2014; Gaudana et al. 2009; Jarvinen and Jarvinen 1996; Rajasekaran et al. 2010).

Microemulsions and Nanoemulsions

Emulsions are fine dispersion of droplets of two immiscible liquids. Nanoemulsions have a particle size in the range of submicron or nanometer range. It comprises of one or more amphiphilic lipids or surfactants. Surfactants are molecule with bipolar structure with hydrophilic and hydrophobic part. Because of their globule sizes, nanoemulsions are thermodynamically unstable dispersions that require high concentration of surfactants to stabilize the formulation which can cause a sticky feeling when the drug is applied to the eye. To make more stable formulations, non-ionic polymers, oxyethylenated or non-oxyethylated sorbitol fatty esters can be used to modify the viscosity of lipid-in-water nanoemulsions without compromising the drug's transparency and bioavailability. The water soluble, non-ionic or neutral polymers of ethylene oxide, vinyl caprolactam, or polyvinyl methyl ether result in nanoemulsion that was stable, transparent with oil globules of less than 100 nm (Lallemand et al. 2012; Hofland et al. 2004; Velagaleti et al. 2010).

The first nanoemulsion approved by the FDA is Restasis® that deliver 0.05 % CsA used for the treatment of dry eye. This drug exhibits satisfactory physicochemical properties, sustained release, higher therapeutic efficacy, and fast onset of action as compared to the solution from Trusopt®. Cyclokat® is another novel nanoemulsion of CsA using Novasorb®, a technology in which cationic nanoemulsion interact well with anionic eye surface, thus improving the bioavailability of this drug to treat dry eye (Chaurasia and Lim 2015; Chauhan and Gulsen 2012; Ako-Adounvo 2014).

Prostaglandins (PGs) are oxygenated cyclic fatty acids that have been used to treat glaucoma. Available PG analogs on the market are bimatoprost, latanoprost, travoprost, and others. However, several of these drugs may have poor water solubility and chemically unstable in aqueous solution. Nanoemulsion formulation made of an oil phase containing PG dispersed in an aqueous phase that is stabilized by a combination of two or more non-ionic surfactants result in a chemically and physically stable emulsion and non-toxic drug such as Catioprost®, a nanoemulsion formulation of latanoprost (Velagaleti et al. 2010).

Semisolid Ophthalmic Forms

In Situ Gels

Also called "Sol-to-Gel" system are viscous liquids that can change from solution to gel by changing pH, temperature and electrolytes, therefore causing slowing drainage from the eye surface and increase the ophthalmic drug bioavailability. Polymers such as anionic gellan gum, poloxamer, cellulose acetate phthalate (CAP) are added to increase the viscosity of the ophthalmic drug gel system when it comes into contact with cation (Rathore and Nema 2009; Gaudana et al. 2009). Combine poloxamer and chitosan allow site-specific drug delivery with gel strength and mucoadhesive properties. Several drugs used in situ gel system that has been approved by the FDA include: pilocarpine, timolol (Timoptic-XE®), tobramycin (Tobradex-ST®), ciprofloxacin, fluconazole, and ganciclovir useful for the treatment of glaucoma, bacterial, fungal or virus infections, respectively. Excipients such as carbopol gels, cellulose derivatives, dextran, gelatin glycerin, polyethylene glycol (PG), and poloxamer 407, and polysorbate 80, polyvinylpyrrolidone are added due to their viscosity enhancing or bioadhesive properties that can significantly improve the ocular retention time (Ako-Adounvo 2014; Velagaleti et al. 2010; Rathore and Nema 2009).

Eye Ointment

Ointments use semisolid or solid hydrocarbon base compound with melting point close to human body temperature. After applying the ointment to the eye, it slowly becomes small drops and stays longer time in conjunctival sac increasing its bioavailability. Although these formulations are well tolerated and safe, they can cause blur vision and irritation to the eye. It should be applied at night time (Baranowski et al. 2014).

Solid Ophthalmic Forms

Soft Contact Lenses Coated with Drug

Soft contact lenses have been used to prevent drug loss, to reduce systemic side effects, and to improve efficacy. The contact lenses served as a matrix for the drug nanoparticles which range from 50 to 200 nm with the lens being clear and not to hinder the vision. Polymers that are widely used with the lenses are cross-linked poly 2-hydroxy ethyl methacrylate and ethylene glycol dimethyl acrylate (HEMA). The drug was released from the lens by diffusion from the particles of the lenses matrix. The lens increased the drug retention time and drug permeability across the cornea, provided sustained drug release and minimizes the systemic absorption *via* nasolacrimal sac (Gong et al. 2009; Lallemand et al. 2012). Drugs such as ciprofloxacin, timolol, dexamethasone, and CsA have been added to these polymer coated lenses successfully, and they stay in the eye for longer period of time than conventional eye drops (Baranowski et al. 2014).

Ocular Inserts

Ocular inserts are solid or semisolid dosage form of drug designed to reside within the ocular cul-de-sac, intended to attach to conjunctiva or directly into the cornea. The insert acts as controlled release reservoir added with polymeric materials such as hydroxyl propyl methyl cellulose (HPMC), hydroxyl ethyl cellulose (HEC), chitosan, or polyvinyl the pyrrolidone (PVP-K-90) and gelatin. The insert causes feeling of foreign body in the eye. Failure of inserting the insert properly into the eye make these forms less desirable. The first ocular insert Ocusert® was the first successful ocular insert to deliver pilocarpine for the treatment of ocular hypertension (Gaudana et al. 2009). Advantages of using ocular insert include: increase bioavailability, precise dosing, controlled release of drugs, avoidance of pulsating drug delivery, minimal systemic absorption, reduced frequency of administration, better sclera or conjunctiva route to the internal target, better shelf life with no need of preservatives (Morrison and Khutoryanskiy 2014). Disadvantages of these formulations include: physical and psychological obstacles of placing solid objects in the eye, vision can interfere due to movement of the inserts, potential accidental loss of the insert, and difficulty to remove the inserts (Baranowski et al. 2014; Aburahna and Mahmoud 2011).

SODIs (Soluble Ophthalmic Drug Inserts)

SODIs are small oval wafers soluble eye inserts that upon application to the conjunctiva sac become moistened by tear fluid; they become soft and adhere to eye surface better. The drug is released in a pulsatile manner to ensure its prolonged effect. The small wafers are made from acrylamide, N-vinylpyrrolidone, and ethyl acrylate. Several drugs such as neomycin, kanamycin, atropine, pilocarpine, dexamethasone, sulfapyridine, and tetracaine had been successfully formulated using SODIs system (Baranowski et al. 2014).

Minidisc Ocular Therapeutic System (OTS)

Minidisc contact lens-sized 4–5 mm in diameter that contains α-ω-bis (4-methacryloxy)-butyl poly (dimethylsiloxane) and poly (hydroxyethyl methacrylate). It is a profiled, convex outside with concave from the inside that increases contact with the eye surface. The OTS can hold hydrophilic or hydrophobic drugs over extended time. The ophthalmic drugs that have been utilized using this system include sulfisoxazole and gentamicin which are able to stay in the sclera from 3–14 days (Nisha and Deepak 2012).

Artificial Tear Inserts

This insert is made with hydroxypropyl cellulose shaped as a long rod. The insert absorbs water from conjunctiva and cornea, forming a hydrophilic layer that stabilizes the tear film and moistens the cornea. Lacrisert® has been marketed for the treatment of dry eye syndrome (Nisha and Deepak 2012).

Collagen Shield

The older version of collagen shields was developed from the porcine sclera, but its use caused discomfort and vision problem. Recent dosage formulation is collasomes, a small piece of collagen ($1 \times 2 \times 0.1$ mm) suspended in 1 % methylcellulose. Collagen shield can carry antibiotics, anti-inflammatory, or other drugs (Tangri and Khurana 2011).

Mini Tablets

Mini tablets are biodegradable, solid drug applied to conjunctiva sac which after long contact with the cornea become gel due to a mucoadhesive polymer, the active ingredients would gradually be released. Mini tablets consist of polymers such as HPMC, HEC, carbopol, chitosan, starch and mannitol excipients and magnesium stearate or sodium stearyl fumarate that has lubricating properties. Active ingredients incorporated into the mini tablets include timolol, piroxicam, ciprofloxacin, gentamicin, and acyclovir (Gaudana et al. 2009; Rajasekaran et al. 2010).

Other Ophthalmic Drug Forms

Sprays

Spray ophthalmic drugs include cycloplegics, mydriatics, and phenylephrine-tropicamide. The distance between the spray device and eye should be around 5–10 cm. The active ingredients in the spray drug should be around 1–4 % which has similar effect like eye drops at a concentration of 1 %. The advantage is the drug can be applied on closed eyelid (Gaudana et al. 2009).

Ocular Iontophoresis

The positively charged molecules are introduced to eye tissue from an anode and the negatively charged ion from the cathode. Iontophoresis enables fast, safe and painless delivery of high concentration of drug to the desired areas (Souza et al. 2013). Drugs delivered using this method include gentamicin, dexamethasone, ciprofloxacin, and ketoconazole.

To deliver drugs to posterior segment, a traditional iontophoretic method is not very effective due to short half-life water soluble compounds that require frequent administration of this drug which is costly and reduce patients' compliances. A Newer approach to this problem is to deliver the active agent and depot forming agent separately. After the depot forming agent was administered, it will make an ionic complex with the active agent and it releases the active drug slowly and in the sustained release into ocular tissues. The depot forming agent such as Ca^{2+}, Sn^{2+}, Fe^{2+}, Fe^{3+}, Mn^{2+}, Zn^{2+}, NH^{4+}, organic anion, chelating agent have been successfully used. This depot should have an opposite charge than the active drug (Gaudana et al. 2012).

Filter Paper Strips

The paper strips sized 5 × 15 mm is covered with 1 mg pigments such as sodium fluorescein. The strips are used as diagnostic tools for the cornea, conjunctiva, palpebral damage, or to check the presence of viral or bacterial eye infections (Tangri and Khurana 2011; Bernard et al. 2012).

Intraocular Delivery

Intraocular drug delivery technique is intended to deposit the drug in the eye directly at the site of action. The advantage of the intraocular delivery is: shorten the distance drugs need to diffuse therefore increasing local concentration, reducing drug delivery to other off-target sites to decrease undesirable side effects, and bypass ocular epithelial and other ocular barriers to increasing bioavailability. The disadvantages are invasive methods, repetitive injections can cause complication such as vitreous hemorrhage, retinal detachment, cataract, endophthalmitis, and inflammation (Gaudana et al. 2012; Jiang et al. 2009). Several methods of intraocular deliveries include: intrastromal, intracameral, suprachoroidal, subretinal and intravitreal.

Intrastromal Delivery

Intrastromal injection is used to deliver macromolecules ophthalmic drugs so the half-life of drugs is extended inside the stroma. Densely packed corneal stromal structure and proteoglycans in the corneal of stroma hinder the diffusion of macromolecule inside the stroma. Anti-vascular endothelium growth factor (VEGF) such as bevacizumab to treat corneal neovascular showed dramatic regression of corneal neovascularization, therefore, increasing visual acuity (Avisar et al. 2010; Storobinsky et al. 2009).

Intracameral Delivery

Intracameral injection is the method of injecting ophthalmic drugs to the anterior segment of the eye. The injection has been used to improve bioavailability to both the anterior and posterior segments of the eye, although this method can't deliver significant concentration to the posterior. Intracameral injection has been used as prophylactic delivery of antibiotic CsA after cataract surgery to prevent endophthalmitis or antifungals that cause deep corneal infections. However to combat the rapid turnover of fluid in the anterior segment, repeated intracameral injections are needed to maintain therapeutic concentrations of the drugs over time, which can cause increased risk of infections (Kim et al. 2014).

Suprachoroidal Delivery

Suprachoroidal are designed to place drugs in the suprachoroidal space (SCS). Normally, the SCS is collapsed due to the deformity of the chorioretinal and the hydrostatic pressure in the eye, but with the positive pressure from the injection, space can be expanded and incorporate fluid. SCS delivery is expected to have a higher bioavailability of the drug, higher concentration of the drug in the choroid, and fewer sides. The SCS injection is best used for sustained release drugs because high blood flow in the chorio-capillaries can wash away small and macromolecules (Kim et al. 2014; Patel et al. 2011).

Subretinal Delivery

The subretinal space is the extracellular space that exists between the photoreceptors of the retina and RPE layer. The subretinal space is a loosely organized space several microns in thickness where macromolecules are specifically injected to the retina. Subretinal space injection has been used clinically, but long-term safety of this injection procedure has not been fully studied (Kim et al. 2014).

Intravitreal Delivery

Intravitreal injection is important to deliver ophthalmic drugs to posterior segment by injecting drugs to all layers of the ocular globe. But with frequent injections, this method is very invasive and can cause retinal damage risks such as retinal detachment, iritis, uveitis, endophthalmitis, intraocular hemorrhage, and cataract. To have their therapeutic effects, the drug molecules must diffuse through the vitreous, chorioretinal, multiple sub-layers of the retina and the RPE to reach the choroid. Drugs injected by this route are cleared either by anterior or posterior route. Several drugs that have been using this delivery include pegaptanib, anti-VEGF bevacizumab and ranibizumab, and aflibercept for treatment of neovascular macular degeneration (Kaur and Kakkar 2014; Kim et al. 2014; Avisar et al. 2010; Donovan 2004).

By formulating the drug into nanoparticles and using intravitreal delivery, it was shown that the drug retention can increase from 1 day to more than 15 days, thus decreasing intravitreal injection frequency (Kaur and Kakkar 2014; Kim et al. 2014; Avisar et al. 2010).

Ocular Implant Delivery

Ocular implants deliver sustained release drugs over several months to years. There are several types of implants: non-biodegradable implants, biodegradable implants, and stimuli-responsive implants. Non-biodegradable implants can trap active drugs by dispersion throughout polymer matrix storage inside a reservoir which is surrounded by a release controlling non-biodegradable polymer membrane. Polyvinyl alcohol (PVA), ethylene vinyl acetate, and silicon are the most commonly used non-biodegradable ocular implants. FDA approved Vitracert® implant containing ganciclovir for the treatment of cytomegalovirus and Retisert® implant containing fluocinolone for the treatment of uveitis, respectively. These implants are sutured to the sclera and need to be inserted and surgically removed. They are effectively implanted in the eye for over 5–8 month for Vitracert® and up to 2.5 years for Retisert® (Yasin 2014; Molokhia et al. 2010).

Biodegradable polymeric implants are made of PLGA, polylactic acid (PLA) which is degradable in the body by water or enzyme to become CO_2 and water. Ozurdex® which contains dexamethasone is FDA's approved for the treatment of diabetic macular edema and is effectively implanted for up to six months in the eye (Yasin 2014; Molokhia et al. 2010).

Figure 5.1. shows classification of ocular implants, Fig. 5.2 shows the location of different implants. Figure 5.3 shows non-biodegradable implant and Fig. 5.4 shows biodegradable implant.

Nano Formulations Delivery

Nano-formulations (NF) use nanoparticles sizes range from 10 to 1000 nm, while microparticles range from 1 to 10 μm. The advantages of nanoparticles or microparticles include: increased corneal penetration, improvement of

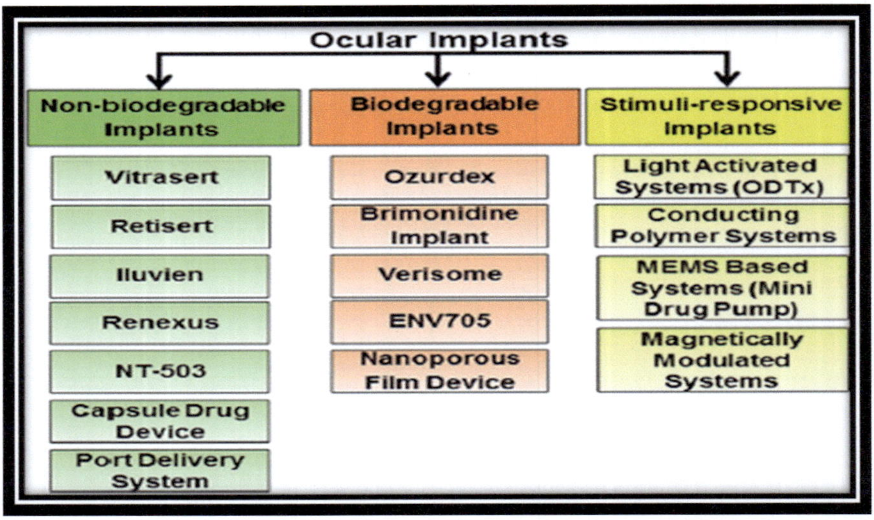

Fig. 5.1 Classification of ocular implant. With permission from Yasin et al. (2014)

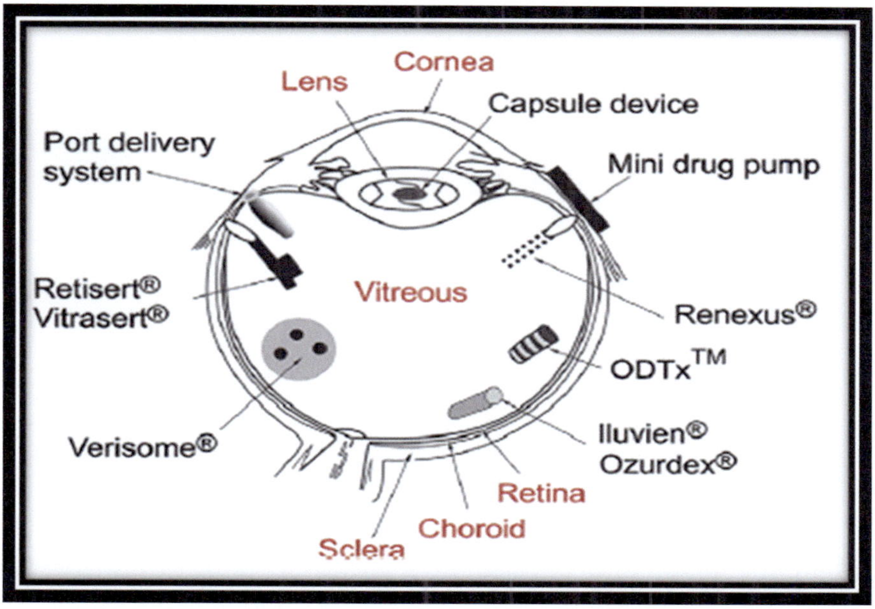

Fig. 5.2 Selected implants and their location in the eye. With permission from Yasin et al. (2014)

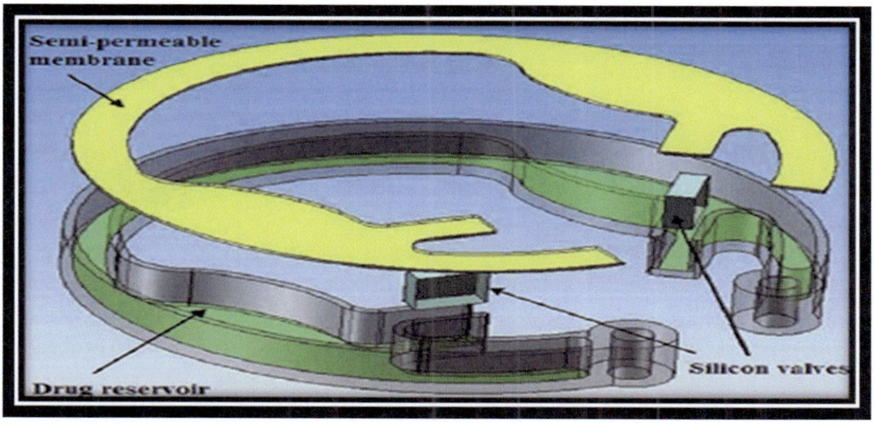

Fig. 5.3 Non-biodegradable implant. Capsule drug device showing the drug reservoir, semipermeable membrane and location of the silicon valves. With permission from Molokhia et al. (2010)

bioavailability of active ingredients. However, it has low capacity to deliver the active drug (Chaurasia et al. 2015).

Nanoparticles usually consist of: nanospheres, solid monolithic spheres built from dense polymer matrix where the active ingredient is scattered, and nanocapsules constituting reservoirs built from polymer membrane surrounding the drug in

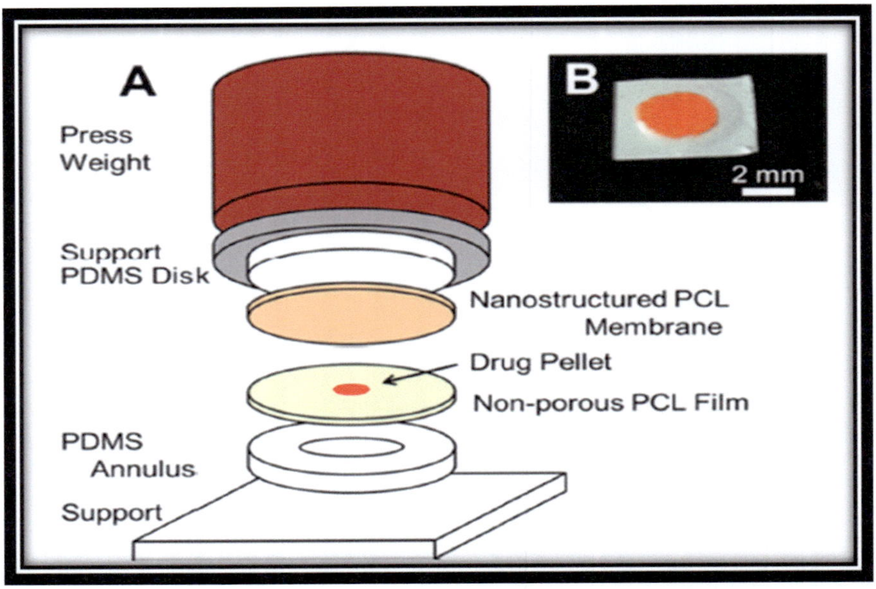

Fig. 5.4 Assembly of nanoporous thin film devices. **a** From the *bottom up*, devices consist of a flat polycaprolactone (PCL) film sandwiched between supporting structures using a press weight. The apparatus containing the constituent device layer is placed on a hot plate to fuse the PCL films, The annulus base support causes the device center to experience considerably less heating. **b** An example of prototypical device loaded with a model protein, fluorescein isothiocyanate labeled bovine serum albumin (FITC-BSA). Reprinted with permission from Bernards DA, Lance KD, Ciaccio NA and Desai TA, 2012. Copyright 2012 American Chemical Society

solid or liquid form. Active ingredients known to deliver by this system include sulfacetamide, levofloxacin, acyclovir, piroxicam, CsA, flurbiprofen, and pilocarpine (Gaudana 2009; Kompella et al. 2013).

Different types of nanoparticles (bioadhesive, sustained release, stealth, targeted and stimuli response) can be used for different routes of administrations as shown in Fig. 5.5.

There are many types of NFs such as: (a) Polymeric nanoparticles; (b) Polymeric nanogels and hydrogels; (c) Liposomes; (d) Nanomicelles; (e) Chitosan nanoparticles; (f) protein nanoparticles; (g) Calcium phosphate nanoparticles; and (h) Dendrimers. Figure 5.6 shows different types of NFs.

Polymeric Nanoparticles

Polymeric nanoparticles such as poly (DL-lactic-*co*-glycolic acid) (PLGA), polyvinyl alcohol, poly (ethylene-co-vinyl acetate), polymethyl methacrylate have been

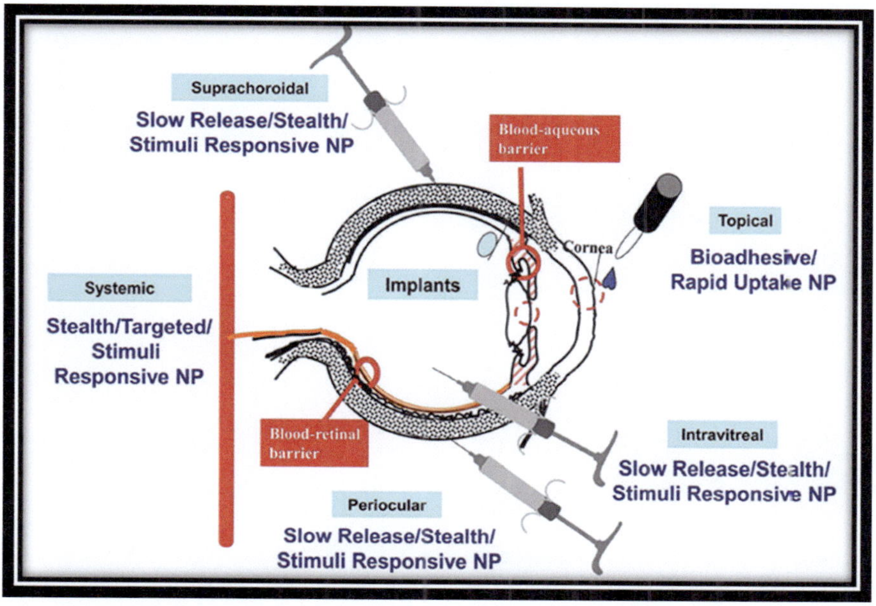

Fig. 5.5 Route of administration (topical, intravitreal, suprachoroidal, periocular and systemic) for delivering different types of nanoparticles (bioadhesive/rapid uptake, sustained release, stealth, targeted, and stimuli response) to the back of the eye. In addition, future applications of nanotechnology include nanofabricated devices, implants, films and particles. With permission from Kompella et al. (2013)

used to carry drugs such as dexamethasone implant (Ozurdex®). These carriers are biodegradable, biocompatible and have minimum side effects (Gong et al. 2009).

Polymeric Nanogels and Hydrogels

Nanogels are hydrogels of a nanoscale network of hydrophilic polymers. Nanogel is loaded with active drug and when the gels collapse, it forms compact nanoparticles. Nanogels can be used for controlled release of hydrophilic and hydrophobic drugs by varying the composition and conformation of polymers and the degree of crosslinking and external stimuli (Lo et al. 2009; Kompella et al. 2013).

Liposomes

Liposomes are vesicular system with diameters ranging from 50 nm to several microns. It contains one or more concentric lipid bilayers separated by aqueous

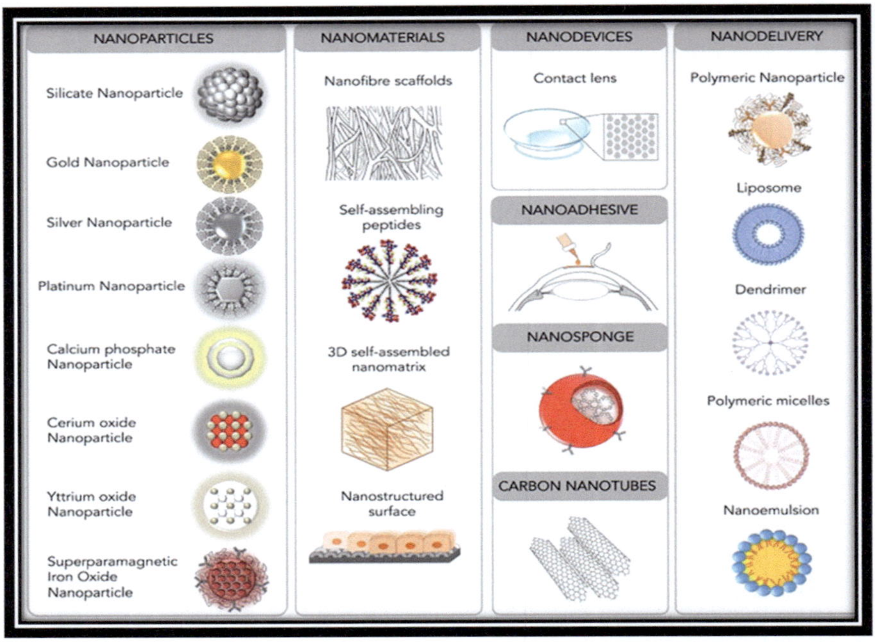

Fig. 5.6 Schematic diagram depicting nanomedicine technique available for the corneal disease. With permission from Chaurasia et al. (2015)

buffer compartment. There is three category of liposomes: (1) multilamellar vesicles (contain more than one lipid bilayer); (2) small unilamellar vesicles (size 10–100 nm), and (3) large unilamellar vesicles (0.1–10 μm).

Liposomes can encapsulate hydrophilic (in aqueous compartment) and hydrophobic (in lipid bilayers) drugs. The advantage of liposomes are biocompatibility, biodegradability, amphiphilic properties, protection from degradation by metabolic enzymes such as protease, esterase, present on the surface of conjunctiva, tear fluid, and corneal surface, low toxicity, have membrane that is flexible and can support deformation stress (Hofland et al. 2004; Tangri and Khurana 2011).

Liposomes are positively charged compounds that disintegrate completely upon contact with high affinity to the negatively charged corneal surface and conjunctival mucoglycoproteins causing slowing elimination of active ingredient from the eye. Liposomes have been successfully used for the treatment of keratitis, uveitis, endophthalmitis and proliferative vitreoretinopathy. Some active ingredients with liposomal form include acyclovir, pilocarpine, acetazolamide, chloramphenicol, and ciprofloxacin. For delivery to the anterior segment, several research endeavor is targeted to coating the exterior surface of liposomes with bioadhesive and penetration enhancing polymers for improving corneal or conjunctival adhesion and permeability (Rothore and Nema 2009; Tangri and Khurana 2011). For posterior segment delivery, research is focused on improving intravitreal drug half-life.

There are several patents with liposomal technology such as use amine derivative lipid component (−NH_2 group is separated from lipid polar head region by carbon-containing spacer-arm), neutral lipid (soybean oil based phospholipid), mucoadhesive agents, carbopol 934P, polyaxomers, carbomers, lectins, cationic lipid consisting stearylamine, cholesterol, dimethyl-dioctadecyl-ammonium bromide, DNA encapsulated liposomes prepared by dilauroyl-phosphatidylcholine (DLPC) or dioleoyl phosphatidylethanolamine (DOPE), and latanoprost-loaded egg-phosphatidylcholine (Jin et al. 2011; Lallemand et al. 2012).

Nanomicelles

Nanomicelles are self-assembling nano-sized (10–100 nm) colloidal dispersion with a hydrophobic core and hydrophilic shell. Nanomicelles solubilize hydrophobic drugs within their hydrophobic core. Some formulations that have been developed using nanomicelles include: rapamycin which had poor aqueous solubility and now has solubility of 1000 folds higher than its origin, or ketorolac for anti-inflammatory using polymeric micelles with copolymer of *N*-isopropyl acrylamide, vinyl pyrrolidone, and acrylic acid cross-link with *N, N*-methylene *bis*acrylamide; and voclosprin a calcineurin inhibitor for the treatment of dry eye use a mixed combination of two non-ionic surfactants, d-α-tocopheryl polyethylene glycol 1000 succinate (vitamin ETPGS) co-stabilized with octyl phenol ethoxylate. Many of these micelles result in the nanometer size range that creates aqueous homogeneous and clear solution that makes nanomicelles ideal ophthalmic drug delivery vehicle (Gong et al. 2009; Chauhan and Gulsen 2012).

Chitosan Nanoparticles

As polysaccharides derived from the natural compound chitin found in the exoskeleton of crustaceans, chitosan provides unique property as mucoadhesive that prolong time on the ocular surface. The hydroxyl groups in the polysaccharide structure of chitosan compete with water on the surface of eye tissue providing additional bio-adhesion. A hybrid of PLGA-chitosan nanoparticles loaded with plasmid delivered to retina after intravitreal injection has shown the drug stayed in the posterior segment longer than without chitosan nanoparticles (Kompella et al. 2013).

Protein Nanoparticles

Albumin, the primary protein of blood plasma, provides advantages over other nanoparticles. Human serum albumin increases drug solubility, decrease drug toxicity, protects drug against oxidation and increase drug half-life. In the eye, serum albumin can pass through BAB. It is the major component of aqueous humor that plays important role in rapidly dividing epithelial lens cells (Kompella et al. 2013).

Calcium Phosphate Nanoparticles

Calcium phosphate (CaP) nanoparticles are inorganic that are chemically stable and biocompatible, works by reducing drug binding to the pigment of the eye, therefore, increase its bioavailability. CaP nanoparticles are biodegradable, dissociates into calcium and phosphate ions. Ophthalmic formulations using 7-hydroxy-2-dipropyl-aminotetralin (7-OH-DPAT) and CaP nanoparticles was shown to decrease the intraocular pressure (IOP) in glaucoma patients compared to using the drug 7-OH-DPAT alone. CaP has delivered methazolamide, a carbonic anhydrase inhibitor for the treatment of glaucoma that has a better-sustained release than free drug without CaP. CaP has also been used as carrier tool for DNA transfection into corneal cells (Malik et al. 2012; Silva et al. 2012; Yavuz et al. 2013).

Dendrimers

Dendrimers are branched, spherical, monodisperse three-dimensional polymer structure of specific size, shape and molecular mass that are typically water soluble. Dendrimers can have neutral, negative, or positive functional group at the terminal. Because the presence of many functional groups on the surface such as $-NH_2$, $-COOH$, and $-OH$ of dendrimers, they can form electrostatic or covalence bonds with the drug. The structure mimic globular proteins, therefore, referred to as "artificial protein". They are used as carriers with the active ingredients inside the polymer structure. Active ingredients used by these systems are: tropicamide, pilocarpine, carteolol, carboplatin, fluocinolone, brimonidine, and timolol (Yavuz et al. 2013).

There are many types of dendrimers such as: (a) poly-amidoamine (PAMAM) dendrimers that are considered ideal carrier for drug delivery due to their high aqueous solubility, large surface groups. Their size ranged from 1.1 to 12.4 nm. Another similar form of PAMAM called poly-amidoamine organosilicon (PAMAMOS) are silicon inverted unimolecular micelles dendrimers that contain exterior hydrophobic organosilicon and interior hydrophilic nucleophilic polyamidoamine; (b) polypropyleneimine (PPI) dendrimers which are toxic due to poly-alkyl amines and usually not a useful dendrimers; (c) poly aryl ether dendrimers which are poor water soluble; and (d) biodegradable dendrimers such as polylysine, polydisulfide amine, polyether, or polyester provide promising candidates as antiviral, antibacterial, and vaccine for ophthalmic drugs. Several dendrimers have been used for treatment of: miosis, mydriasis, glaucoma, conjunctivitis, intraocular infections, retinoblastoma, ocular hypertension, retinal neuroinflammation, cataract incisions. Dendrimers delivery system have been useful to increase corneal residence time, to prolong reduction of IOP, to reduce toxicity, to enhance corneal transport, to increase antimicrobial activity, to increase half-life and bioavailability, to promote adhesion and proliferation of human corneal epithelial cells, to increase uptake of drugs, and to expedite wound healing (Malik et al. 2012; Silva et al. 2012).

Dendrimers have the ability to encapsulate drug molecules into their internal cavities to enhance drug solubility, permeability and retention time in the eye. Drug absorption is much better for cationic molecule; followed by uncharged and the least absorption is anionic molecules. Cationic dendrimers have increased permeability due to their interacting with lipid bilayers but it is also more toxic. In order to reduce toxicity, it is necessary to modify the surface with amine group that makes the dendrimer to be neutral or have anionic moieties (Malik et al. 2012; Silva et al. 2012; Yavuz et al. 2013).

Niosomes and Discosomes

Niosomes are chemically stable non-ionic surfactants Solulan C24, two-layered carriers used for hydrophilic or hydrophobic drug. These biodegradable, biocompatible, and non-immunogenic carriers extend the contact period between the drug and cornea for better drug bioavailability.

Discosomes are modified from the of niosomes with disc shape that fits better into conjunctival sac and do not enter the general circulation. Active ingredients known to use these systems are ganciclovir, timolol, cyclopentolate (Cholkar et al. 2013).

NFs can be administered using various routes such as intravenous, topical, periocular, intravitreal or suprachoroidal. The selection of route of administration depends on drug properties, disease state, target tissues/cell, and nano material properties such drug loading, encapsulation, efficiency, and safety. Various nanocarriers ranging from the more established systems like liposomes, polymeric nanoparticles, solid lipid nanoparticles, Nano micelles, to the most novel systems like Nanosuspensions, Nanoemulsions, nanocrystals, hydrogel, and dendrimers. NF with the whole antibody, Fab nanosphere has been shown to increase retention time following intravitreal administration from 1 day to 28 days (Cholkar et al. 2013).

However, several aspects need to be addressed to produce successful nano delivery such as minimizing toxic effects, low quantities of impurities, batch to batch reproducibility and safety in large-scale manufacturing should be established. Nanoparticles can aggregate and grow in size during storage and important approach to minimize aggregation in aqueous, biological media, and shelf, the tendency of nanoparticles to aggregate when administered in periocular space, the need to influence vitreous humor, aqueous humor, blood with nanoparticles properties.

Summary

Despite many achievements in the field of ophthalmic delivery, majority of active substance delivered to the eye are in the forms of topical such as eye drops, ophthalmic solutions. Intraocular delivery is intended to deposit the drug in the eye directly at the site of action, but these methods are invasive. Scientists are still looking for the "perfect" ophthalmic system that would possess desired properties such as controlled release, minimizing systemic effects, ease of use, and extended retention time at the site of application.

References

Aburahma MH, Mahmoud AA (2011) Biodegradable ocular inserts for sustained delivery of brimonidine tartrate: preparation and in vitro/in vivo evaluation. AAPS Pharm Sci Tech 12 (4):1335–1347

Ako-Adounvo AM, Nagarwal RC, Oliveira L et al (2014) Recent patents on ophthalmic nanoformulations and therapeutic implications. Recent Pat Drug Deliv Formul 8(3):193–201

Arepalli S, Kaliki S, Shields CL (2015) Choroidal metastases: origin, features, and therapy. Indian J Opthalmology 63:122–127

Avisar I, Weinberger D, Kremer I (2010) Effect of subconjunctival and intraocular bevacizumab injections on corneal neovascularization in a mouse model. Curr Eye Res 35:108–115

Baranowski P, Karolewicz B, Gajda M, Pluta J (2014) Ophthalmic drug dosage forms: characterization and research methods. Sci World J doi:10.1155/2014/861904

Bell STD, Chu T, He Q, Potter DE (2004) Intraocular delivery compositions and methods. WO2004050065A1

Bernards DA, Bhisitkul RB, Desai T (2014) Zero-order sustained drug delivery to the retina from a nanoporous film device. Drug Deliv 48:20–21

Chauhan A, Gulsen D (2012) Ophthalmic drug delivery system. US8273366

Chaurasia SS, Lim RR, Lakshminarayanan R, Mohan RR (2015) Nanomedicine approaches for cornea diseases. J Funct Biomater 6:277–298

Chemuturi NV, Yanez JA (2013) The role of xenobiotic transporters in ophthalmic drug delivery. J Pharm Sci 16:683–707

Cholkar K, Patel SP, Vadlapudi AD, Mitra AK (2013) Novel strategies for anterior segment ocular drug delivery. J Ocu Pharm Ther 29(2):106–123

Donovan S (2004) Intravitreal botulinum toxin implant. US200475871

Gaudana RJ, Jwala J, Boddu SHS, Mitra AK (2009) A Recent perspective in ocular drug delivery. Pharm Res 26(5):1197–1216

Gaudana RJ, Gokulgandhi MR, Boddu SHS, Mitra AK (2012) Recent overview of ocular patents. Recent Pat Drug Deliv Formul 6(2):95–106

Gong X, Peng S, Wen W et al (2009) Design and fabrication of magnetically functionalized core/shellmicrospheres for smart drug delivery. Adv Funct Mater 19:292–297

Hofland H, Bongianni J, Wheeler T (2004) Ophthalmic liposome compositions and uses thereof. US20040224010

Jarvinen T, Jarvinen K (1996) Prodrugs for improved ocular drug delivery. Adv Drug Del Rev 19 (2):203–224

Jiang J, Moore JS, Edelhauser HF, Prausnitz MR (2009) Intrascleral drug delivery to the eye using hollow microneedles. Pharm Res 26:395–403

Jin J, Zhou KK, Oark K et al (2011) Anti-inflammatory and anti-angiogenic effects of nanoparticles mediated delivery of a natural angiogenic inhibitor. Invest Opthalmol Vis Sci 52:6230–6237

Kaur IP, Kakkar S (2014) Nanotherapy for posterior eye diseases. J Control Rel 193:100–112

Kim YC, Chiang B, Wu X, Prausnitz MR (2014) Ocular delivery of macromolecules. J Controlled Release 190:172–181

Kompella UB, Amrite AC, Ravi RP, Durazo SA (2013) Nanomedicines for back of the eye drug delivery, gene delivery, and imaging. Prog Retin Eye Res 36:172–198

Lallemand F, Daull P, Benita S et al (2012) Successfully improving ocular drug delivery using the cationic nanoemulsion. Novasorb J Drug Deliv 604204

Lo R, Li PY, Saati S et al (2009) A passive MEMS drug delivery pump for treatment of ocular diseases. Biomed Microdevices 11:959–970

Malik AS, Chaudhary S, Garg G, Tomar A (2012) Dendrimers: a tool for drug delivery. Adv Biological Res 6(4):165–169

Martin DF, Maguire MG, Fine GS et al (2012) Ranibizumab and bevacizumab for treatment of neovascular age-related macular degeneration: two-year results. Opthalmology 119:1388–1398

Molokhia SA, Sant H, Simonis J et al (2010) The capsule drug device: novel approach for drug delivery to the eye. Vis Res 50:680–685

Morrison PWJ, Khutoryanskiy VV (2014) Advances in ophthalmic drug delivery. Ther Deliv 5 (12):1297–1315

Nisha S, Deepak K (2012) An insight to ophthalmic drug delivery system. Inter J Pharma Studies Res 3(2):9–13

Patel SR, Lin AS, Edelhauser HF, Prausnitz MR (2011) Suprachoroidal drug delivery to the back of the eye using hollow microneedles. Pharm Res 28:166–176

Rajasekaran A, Kumaran KSGA, Preetha JP, Karthika K (2010) A comparative review on conventional and advanced ocular drug delivery formulations. Int J Pharm Tech Res 2(1):668–674

Rathore KS, Nema RK (2009) An insight into ophthalmic drug delivery system. Int J Pharm Sci Drug Res 1(1):1–5

Silva NP, Menacho FP, Chorilli M (2012) Dendrimers as potential platform in nanotechnology-based drug delivery systems. IOSR J Pharm 2(5):23–30

Souza JG, Dias K, Pereira TA et al (2013) Topical delivery of ocular therapeutics: carrier systems and physical methods. J Pharm Pharmacol 66:507–530

Storobinsky DO, Lubin BC, Hasanreisoglu M, Goldenberg-Cohen N (2009) Effect of subconjunctival and intraocular bevacizumab injection on angiogenic gene expression levels in mouse model of corneal neovascularization. Mol Vis 15:2326–2338

Tangri P, Khurana S (2011) Basics of ocular drug delivery system. Int J Res Pharm Biomed Sci 2 (4):1541–1552

Velagaleti PR, Anglade E, Khan IJ et al (2010) Topical delivery of hydrophobic drugs using a novel mixed nanomicelles technology to treat diseases of the anterior and posterior segments of the eye. Drug Deliv Technol 10(4):42–47

Yasin MN, Svirskis D, Seyfoddin A, Rupenthal ID (2014) Implants for drug delivery to the posterior segment of the eye: A focus on stimuli-responsive and tunable release systems. J Controlled Release 196:208–221

Yavuz B, Pehlivan SB, Unlu N (2013) Dendrimeric system and their applications in ocular drug delivery. Sci World J. doi:10.1155/2013/732340

Chapter 6
Nanotechnology for Transcorneal Drug Targeting in Glaucoma: Challenges and Progress

Ameeduzzafar Zafar, Javed Ahmad, Sohail Akhter and Richard T. Addo

Abstract The eye is a highly protected organ, and designing ocular formulation for effective therapy, is challenging for drug delivery researcher. The anatomical and physiological barriers resulted in a low ocular bioavailability of administered drugs. Poor bioavailability of ocularly administered drugs is mainly due to factors responsible for precorneal loss (like tear dynamics, non-productive absorption, a transient residence time in the cul-de-sac, and relative impermeability of the corneal epithelial membrane). Due to these constraints, less than 5 % of the administered dose is absorbed from the conventional ophthalmic dosage forms. Vision-threatening diseases like glaucoma alter the physiology and molecular mechanism of vision. Ocular drug delivery in this dreadful condition is quite challenging. Though, the potential use of a nanoparticulate system as drug carriers has led to the development of many different colloidal delivery vehicles for targeted delivery in glaucoma. Drug loaded colloidal carriers associated with several favorable biological characteristics such as biodegradability, biocompatibility and mucoadhesiveness have been found to be effective in transcorneal drug targeting in glaucoma. These nanoparticulate systems exhibited better ocular drug efficacy by improving ocular bioavailability without blurring the vision in glaucoma. This

A. Zafar
Department of Pharmaceutics, Faculty of Pharmacy, Jamia Hamdard University, New Delhi 110062, India

J. Ahmad
Department of Pharmaceutics, National Institute of Pharmaceutical Education and Research (NIPER), Raebareli, Uttar Pradesh 229010, India

S. Akhter
LE STUDIUM® Loire Valley Institute for Advanced Studies, Centre-Val de Loire, France

S. Akhter
Nucleic Acids Transfer by Non-Viral Methods, Centre de Biophysique Moléculaire, CNRS, UPR4301 Orléans, France

R.T. Addo (✉)
Department of Pharmaceutical Sciences, School of Pharmacy, Union University, Jackson, TN 38305, USA
e-mail: raddo@uu.edu

R.T. Addo (ed.), *Ocular Drug Delivery: Advances, Challenges and Applications*,
DOI 10.1007/978-3-319-47691-9_6

chapter aims to briefly discuss the ocular barriers to glaucoma drug delivery along with nanotechnology mediated transcorneal drug targeting.

Keywords Glaucoma · Bioavailability · Nanotechnology · Colloidal carrier · Transcorneal drug targeting

Introduction

Glaucoma is a group of eye sicknesses that is associated with damage to optic nerve. The optic nerve is the main nerve of the eye (posterior part of the eye), which are responsible for transmitting electrical impulses to the brain. It is also characterized by a progressive form of optic nerve damage associated with raised (>21 mmHg) intraocular tension and reduction of the visual field (Zimmerman 1993; Jarvinen et al. 2000; Langman et al. 2005; Quaranta et al. 2013a, b; Hyun et al. 2013). It occurs when an imbalance in production and drainage of ocular fluid (aqueous humor) that resulted in increased eye pressure (Van Buskirk and Cioffi 1992; Duijm et al. 1997; Casson et al. 2012; Rhee 2013). Usually patients with glaucoma, especially those suffering from normal tension glaucoma (NTG) or progressive high tension glaucoma (HTG), have slower blood flow velocity in the retina, (Langman et al. 2005) in the choroid (Duijm et al. 1997) and in the optic nerve head (Michelson et al. 1996). Blood flow is also reduced in the retrobulbar vessels, (Satilmis et al. 2003; Doina et al. 2004) carotid arteries, and particularly in the peripheral capillaries. In addition, it has also been observed that patients with glaucoma frequently suffer from ischemic lesions in other parts of the body such as the brain (Stroman et al. 1995) ear (Susanna and Basseto 1992) and heart. Glaucoma-like optic nerve head excavations have also been demonstrated in animals following induced ischemia. These reports suggest that the vascular mechanism of damage in glaucoma is at least in some respects a primary causative factor. Approximately 70 million people around the world are affected and estimated 7.5 million blind worldwide resulting from glaucoma. In India, at least 12 million peoples are affected with glaucoma. The disease is more common in old age, however approximately 2–3 persons per hundred are affected above the age of 40 years (Quigley 1996; Quigley and Broman 2006; Kowing et al. 2010; Jain et al. 2014).

Types of Glaucoma

There are many types of glaucoma, but the two most common types i.e., open-angle glaucoma and closed angle (angle closure) glaucoma.

Primary Open-Angle Glaucoma

Primary open-angle glaucoma (POAG) a chronic, slowly progressive optic neuropathy, characterized by progressive excavation of the optic nerve head and a distinctive pattern of visual field defects. It is about 90 % of all glaucoma cases. The disease is multifactorial in origin and is associated more closely with elevated intraocular pressure (IOP) resulting from reduced drainage of aqueous humor because trabecular meshwork becomes blocked and fluid cannot be transported to the normal drainage canals (Van Buskirk and Cioffi 1992; Piltz-Seymour et al. 2001; Doina et al. 2004; Gherghel et al. 2004; Ohtake et al. 2004).

Closed-Angle Glaucoma

It is also called acute glaucoma or angle closure glaucoma, accounts for about 9 % of all glaucoma cases and occurs when the opening between cornea and iris get narrows, such that the fluid cannot get to the trabecular meshwork and normal drainage channels (Fig. 6.1) (Shields 1992).

There are various anti-glaucoma drugs (Nathanson 1980; De-Santis 1994; Saxena et al. 2002; Tataru and Purcarea 2012) used for the treatment and available on the market, shown in Table 6.1. It acts on the aqueous humor dynamics to reduce the intraocular pressure mainly by three mechanisms.

- Decrease aqueous production in the ciliary body
- Increase aqueous humor outflow through the trabecular meshwork and
- Increase aqueous humor outflow via the uvea-scleral pathway.

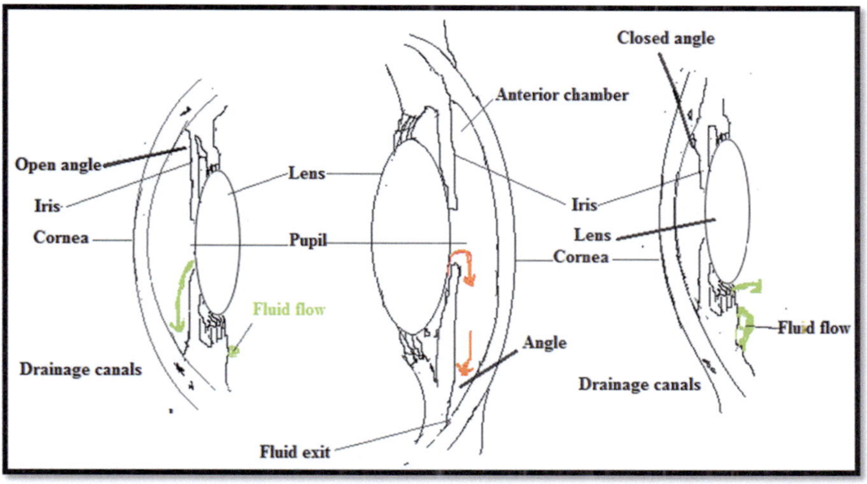

Fig. 6.1 Types of glaucoma

Table 6.1 Different drugs used in treatment of glaucoma

Drug	Dosage form	Action
Cholinergic agonist		
Pilocarpine	Eye drops (0.5–4 %)	It increases the trabecular outflow due to ciliary body contraction
Adrenergic agonists		
Dipivefrin (prodrug of epinephrine)	Eye drops (0.1 %)	It increases aqueous humor formation in the early phase presumably due to its α-adrenergic effect
		It also increases trabecular outflow probably by stimulating β2-adrenergic receptors in the trabecular meshwork
Epinephrine	Eye drops (0.25–2 %)	It decreases aqueous humor formation in the early phase presumably due to its α-adrenergic effect
		It also increases trabecular outflow probably by stimulating β2-adrenergic receptors in the trabecular meshwork
α_2-adrenergic agonist		
Apraclonidine	Eye drops (0.5–1 %)	Increasing trabecular meshwork outflow by reducing episcleral venous pressure and may also increase uvea-scleral outflow by an increase in prostaglandin synthesis
Brimonidine	Eye drops (0.2 %)	
Adrenergic antagonists		
Selective β1-blockers		
Betaxolol	Eye drops (0.5 %)	They act solely by reducing the aqueous humor production
Atenolol	Eye drops (4 %)	They block the β-receptors in the iris and ciliary body and thereby cause a significant reduction in IOP
Metoprolol	Eye drops (1–4 %)	
Non-selective β-Blockers		
Timolol,	Eye drops (0.25–0.5 %)	It reduces intraocular pressure by reducing Aqueous humor formation as well as by enhancing the outflow facility
Levobunolol	0.5 % eye drops	
Carteolol	1–2 % eye drops	
Carbonic anhydrase inhibitors		
Acetazolamide	250 mg tablets	It reversibly blocks the enzyme carbonic anhydrase in the ciliary body and thus suppresses aqueous humor production
Methazolamide	50 mg tablets	
Dorzolamide	Eye drops (2 %)	It penetrates cornea, inhibits carbonic anhydrase-II in the ciliary body, slows the production of local bicarbonates and thus decreases sodium and fluid transport which in turn reduces the secretion of aqueous humor
Brinzolamide	Eye drops (1 %)	
Prostaglandin analogs		
Latanoprost	Eye drops (0.005 %)	It undergoes enzymatic hydrolysis in cornea and gets activated by the acid of latanoprost
Travoprost	Eye drops (0.004 %)	It acts by enhancing uvea-scleral outflow rather than altering the conventional trabeculo-canalicular aqueous outflow
Unoprostone	Solution (0.12–0.15 %)	It acts by enhancing uvea-scleral outflow without affecting aqueous humor production

Ocular Drug Delivery in Glaucoma: Challenges and Opportunity

The development of drug delivery approach for the transportation of drug in a bio-available and safe manner to the target site is now becoming an exceedingly important area of biopharmaceutical researches.

There are various types of novel drug delivery techniques in every year and every part of the body has been attempted to deliver the active constituent to achieve the maximum bioavailability and minimum side effect as a potential target for the site of action. As a result, different smart drug delivery technologies with considerable outcomes have been reported for BCS class-II (low soluble) and class-IV (low soluble) drugs, proteins, peptides, herbal drugs, etc. among the novel techniques i.e., vesicular system, in situ gel system (temp, ion, pH and temp-ion dependent), bioadhesive, lipid system (solid lipid nanoparticles and nanostructured lipid carrier) nanotubes, dendrimers, emulsion (nanoemulsion, Microemulsion) and polymeric nanoparticles (biodegradable and non-biodegradable), implant, inserts etc. are currently under intensive exploratory studies (Akhter et al. 2011). Apart from that, drug delivery to the ophthalmic drug delivery has remained as one of the most challenging research for scientists, requires a series of specified characteristics on every step during the development of formulation according to the physiological and anatomical structure of the eye because it interferes with the fate of the administered drug and bioactive. To treat the local ophthalmic diseases, liquid eye drop (conventional) is the most considerable and desirable dosage forms for the treatment of most of the ocular diseases due to their ease of administration and clinical compliance of the patients. This conventional dosage form accounts for nearly about 90 % of dosage forms available in the market owing to their simplicity and good acceptance by patients. However, conventional eye drops, most of which are present in the solution form usually have quite a limited therapeutic efficacy due to low bioavailability (Jain et al. 2011). The conventional eye drop have some drawbacks (1) frequent instillation is required, to get the expected therapeutic effect, and this leads to growing inconvenience and adverse effects (Jain et al. 2011). (2) Only less than 5 % administered dose reaches into the active site. (3) The precorneal drug loss is due to blinking and high tear fluid turnover, which may pass into systemic circulation through nasolacrimal duct and conjunctiva (Bourlais et al. 1998). The objective of using the novel delivery system is based on these considerations thus enhancing the precorneal residence time, slow release of drug from a dosage form, slow removal of a drug from the therapeutic site and improvement of both intracellular and paracellular pathways of epithelial cells. It would be of enormous benefit to reduce the dose, dosing frequency and consequently, reduce the local side-effects and systemic side effect inborn to applied drugs. So, drug delivery in ocular therapeutics is a challenging concern and is a subject of interest to scientists working in the multidisciplinary areas pertaining to the eye (Akhter et al. 2011). In case of ocular drug delivery system most of the previous published research data on novel techniques (nano formulation) suggests many considerable point related to formulation i.e., an appropriate particle size and a narrow size range (<200 nm), adequate bioavailability and compatibility with ocular tissues, ensuring low irritation, high residence time, and ensuring low systemic side effect, should be sought for every suspended drug (Sahoo et al. 2008). Thus, an optimum delivery system should be the one which can be delivered in the form of eye drops, causing no blurred vision or irritability and would need to be applied with low dosing frequency and better patient compliance. Despite numerous scientific efforts, efficient ocular drug delivery remains a challenge for the pharmaceutical scientists. Consequently, the design of a system with improved drug delivery properties to the ocular surface would be a promising step towards the management of external ocular diseases. Ophthalmic drug delivery, probably more than any other route of administration, may benefit from the uniqueness of nanotechnology-based drug delivery (Sahoo and Labhasetwar 2003). The use of nanocarriers provides attractive replacements for topical ocular drug delivery, mainly because of their capacity to protect the encapsulated molecule, along with its facilitated transport to the different compartments of the eye (Losa et al. 1993; Kayseri et al. 2005).

Additionally, nanoparticulate formulation may offer the possibility of controlling drug delivery (release of drug from formulation), thus being attractive vehicles for the treatment of some chronic ocular diseases like glaucoma

Anatomy and Physiology of Human Eye

The eye is characterized by its complex structure as well as high resistance to foreign substances including drugs. The anterior and posterior segments of the eye, although in juxtaposition to each other, and very different in their anatomical and physiological aspects, function both independently and in tandem upon application of an ocular preparation. While it has been known for long that conventional topical formulations are amenable to application to the anterior portion, most of the applied dose is lost due to the defensive mechanism of the eye. Consequently, a much concerted effort has been directed towards increased retention of the applied dose on the eye surface, with the hope that such increased retention will result in better therapeutic effect and lowered local and/or systemic effects. Since most drugs poorly penetrate the cornea, fulminating diseases of the posterior segment viz. vitreous, retina and choroid are required to be treated with either systemic administration or through intravitreal injections and vitreal implants. While therapy with systemic administration requires large doses due to strong blood-ocular tissue barrier, the other two routes are very invasive requiring skilled administration and are associated with a high degree of risk, such as the development of retinal detachment and endophthalmitis. Clearly, there is a strong case in favor of formulating ocular delivery systems by focusing on improved ocular bioavailability and extended drug effect in targeted tissues. Prolonging pre-corneal residence time through viscosity enhancers and gels has only a limited value because such liquid formulations are eliminated by the usual routes in the ocular domain. The selection of enhancer to maximize drug transport requires great caution, because corneal/conjunctival tissues are highly sensitive towards penetration enhancers. An alternative approach is to develop a drug delivery system that would circumvent the problems associated with the conventional systems, and provide the advantages of targeted delivery of drugs for an extended period of time and be patient-friendly. The latter requisite becomes more crucial in cases where the patient has to use the drug preparation throughout his life, e.g. in glaucoma. These advantages have been reported in the literature through the use of nanoparticles.

Barriers in Ocular Drug Delivery

The different barriers in ocular drug delivery are discussed below.

Drug Loss from the Ocular Surface

After instillation, the flow of lacrimal fluid removes instilled formulation from the surface of the eye. Even though the lacrimal turnover rate is only about 1/min the excess volume of the instilled fluid is flown to the nasolacrimal duct rapidly in a couple of minutes (Schoenwald 1990). Another source of non-productive drug removal is its systemic absorption instead of ocular absorption. The systemic absorption of a drug from any dosage form may take place through two ways, first absorption directly from the conjunctival sac via local blood capillaries secondly by nasolacrimal duct or cavity (Ananthula et al. 2009). The systemic absorption of the drug through topical ocular routes depends on upon the particles size or molecular weight of the drug. The smaller particle size or very small molecular weight drug are absorbed into systemic circulation rapidly within few minutes, consequently decreasing the concentration of drug in lacrimal fluid extensively. This leads to lower bioavailability of drug i.e., less than 5 % (Schoenwald 1990; Lee and Robinson 1986). So the novel drug delivery has the capability to resist this loss of drug, which has been proved by many scientists.

Lacrimal Fluid-Eye Barriers

When the formulation is instilled in the eye surface it passes into the lacrimal fluid, then cornea. The drug absorption through the corneal epithelium is limited or lower because of the presence of the lacrimal fluid barrier. The corneal barrier is formed upon maturation of the epithelial cells. They migrate from the limbal region towards the center of the cornea and to the apical surface. The most apical corneal epithelial cells form tight junctions that limit the paracellular drug permeation. Therefore, the lipophilic drugs cross in high magnitude as compared to a hydrophilic drug because proximal part of corneal epithelium has more permeability for a lipophilic drug

(Machaand and Mitra 2003). Due to the tightness of the corneal epithelial layer, transcorneal permeation is the main route of drug doorway from the lacrimal fluid to the aqueous humor. In general, the conjunctiva is more permeable epithelium than the corneal epithelium and its surface area approx 20 times greater than that of the corneal epithelium. Drug absorption across the bulbar conjunctiva has gained increasing attention recently since conjunctiva is also quite permeable to the hydrophilic and large molecules weight or large particle size. Therefore, it may serve as a route of absorption for larger bio-organic compounds such as proteins and peptides. Clinically drugs used are generally small and fairly lipophilic. Thus, the corneal route is currently dominating as compared to other ocular routes for the treatment of glaucoma (Satilmis et al. 2003).

Blood-Ocular Barriers

The eye is a vital organ of the body which is protected from the xenobiotics in the blood stream by blood-ocular barriers. These barriers have two parts: blood-aqueous barrier and blood-retina barrier (BRB) which is shown in Fig. 6.2. The anterior blood-eye barrier is composed of the endothelial cells in the uvea. The anterior blood-eye barrier prevents the admittance of plasma albumin into the aqueous humor, as well as limits access of hydrophilic drugs from plasma into the aqueous humor. Inflammation may disrupt the integrity of this barrier causing unlimited drug distribution to the anterior chamber. In fact, the permeability of this barrier is poorly characterized for a hydrophilic drug. This barrier comprised of retinal pigment epithelium (RPE, inner BRB) and the tight walls of retinal blood capillaries (outer BRB, Langman et al. 2005; Quaranta et al. 2013a, b). There are no diffusional barriers between the extracellular fluid (ECF) of the retina and the adjacent vitreous, nor does the vitreous body itself extensively obstruct the diffusion of substances. Principally, the drug molecules may move in the vitreous by two

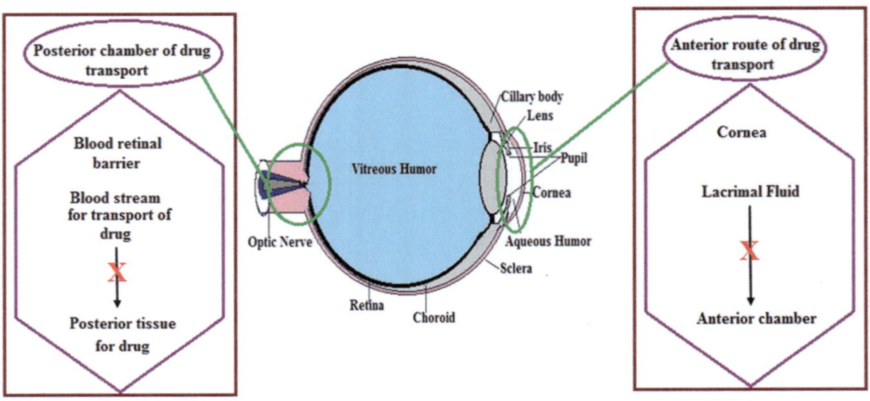

Fig. 6.2 Routes and barriers of eye

different mechanisms: diffusion or bulk flow. Unlike retinal capillaries, the vasculature of the choroid has extensive blood flow and leaky (permeable) walls. The drug molecules are easily interred to the choroidal extravascular space and distributed to the retina. But the distribution of drug molecules into the retina is limited by the RPE (inner BRB) and retinal endothelial cell (outer BRB) due to the tight junction (analog to brain vessels). The choroidal is highly vascularised but it constitutes only a small fraction of the entire blood flow in the body. Thus, without specific targeting systems only a small fraction drug of the oral or intravenous dose reached the retina and choroid (Maurice and Mishima 1984; Hornof et al. 2005). As per many researchers the outer and inner components of the BRB, having different kind of absorptive transport processes which are capable of removing potentially harmful substances from the extracellular fluid of the retina and vitreous (Rapoport 1976).

Drug Targeting Through Topical Transcorneal Route

Topical transcorneal route of drug administration was found to be a potential strategy in the management of glaucoma. Drug penetration through transcorneal route is mainly affected by the corneal barrier, physicochemical characteristics of the drugs as well as active ion transport system present at the cornea.

Anatomy and Physiology of Transcornea Route

The cornea is an optically transparent tissue. It acts as the principal refractive element of the eye. The diameter of cornea approx 11.7 mm with a radius of curvature of the anterior surface of about 7.8 mm (Sahoo et al. 2008). The thickness of the cornea is about 0.5–0.7 mm and it is thicker in the center than in the limbus (Urtti et al. 1990, 1994). The cornea is composed of epithelium, Bowman's membrane, stroma, Descement's membrane and endothelium. The relative thicknesses of corneal epithelium (50–90 μm), stroma and endothelium are about 0.1: 1.0:0.01 (Maurice and Mishima 1984). Usually, the corneal epithelium is the main barrier of drug absorption into the eye (Jarvinen et al. 2000). Compared to many other epithelial tissues (intestinal, nasal, bronchial, tracheal), corneal epithelium is fairly impermeable, but it is more permeable compared to the stratum corneum of the skin (Liaw and Robinson 1992). The stratified corneal epithelium acts as a protective barrier to invasion of foreign substances and also as a barrier to ion transport. The corneal epithelium consists of a basal layer of columnar cells, two to three layers of wing cells and one or two outermost layers of squamous, polygonal shaped superficial cells (Maurice and Mishima 1984). The surface areas of superficial squamous cells of the rabbit corneal epithelium at central, paracentral and peripheral sites vary substantially. Cell divisions occur in the basal layer of epithelium and cellular differentiation occurs gradually as cells move towards the corneal surface. In a healthy corneal epithelium, the intercellular tight junctions (zonula occludens) completely surround the most superficial cells, but the intercellular spaces are wider between wing cells and between basal cells (Maurice and Mishima 1984). This allows paracellular diffusion of large molecules like horseradish peroxidase (MW 40,000) through these layers of cells only (Hämäläinen et al. 1997). When the superficial cell layers of the epithelium in excised pigmented rabbit cornea was first devitalized, gradual formation of tight junctions between the epithelial cells was observed within 8 h (Geroski and Edelhauser 2001) and the complete reformation of the corneal epithelial barrier took place in 5 days following removal of the corneal epithelium of pigmented rabbits (Newell 1986). Perfusion studies have been done in an attempt to estimate the size range of the intercellular space in the epithelium. Small hydrophilic molecules, such as glycerol (MW 92, 0.6 nm) (Maurice and Mishima 1984) and polyethylene glycols 200 and 400 (Huang et al. 1989) are able to penetrate through intercellular spaces of the corneal epithelium, but inulin (MW 5000, 1.5 nm) (Maurice and Mishima 1984) and horseradish peroxidase (MW 40,000 about 3 nm) (Hämäläinen et al. 1997) molecules are too large for paracellular diffusion across the corneal epithelium. Corneal and conjunctival permeability coefficients of hydrophilic P-blockers with similar molecular weights (about 300) but different partition behaviors (octanol/buffer partition coefficients varied from −0.62 to 0.93) suggesting a paracellular penetration route into the eye. The stroma of the cornea is highly hydrophilic in nature because it contains mostly water molecule. Human stroma is composed mainly of collagen fibrils with a uniform diameter of 25–35 nm which runs parallel to each other to form collagen bundles with varying widths and thicknesses. The regular arrangement of collagen fibrils in collagen bundles is important for normal visual acuity. The corneal stroma constitute with corneal fibroblasts (keratocytes, major cellular component) that occupy 2–3 % of the total volume of the corneal stroma. Due to their comparatively loose structure, the drugs having molecular or particles size up to 500,000 can diffuse in the normal stroma (Jarvinen et al. 2000). But it is rate limiting barrier of ocular absorption for most lipophilic drugs (Langman et al. 2005). This is not due to the physical barrier of the stroma, but rather to the low partitioning between lipophilic compounds of lipoid epithelium and hydrophilic stroma. The corneal endothelium is accountable for maintaining normal corneal hydration. The corneal endothelium is a single layer of hexagonal cells covering the posterior surface of the cornea (Maurice and Mishima 1984). Junctional complexes between endothelial cells were seen in the New Zealand White rabbit from 13 days after birth (Kaye et al. 1973). Tight junctions are present in corneal endothelium but they are not as tight as those in epithelium as judged by its permeability to horseradish peroxidase (Liaw and Robinson 1993). It has been expected that drug molecules with dimensions up to approx 20 nm can diffuse across normal endothelium cell (Gaudana et al. 2009a, b).

Ocular Drug Transporter

The conventional approach to improving ocular bioavailability of drug molecules exploited some chemical modification to accomplish the desired solubility and lipophilicity characteristics in the drug molecule. But this approach is very complicated and expensive. However, a more rational approach would be a transporter-targeted modification of the drug i.e., drug molecule bound with a transporter. Transporters are membrane-bound proteins that play a vital role in the active transport of nutrients and drug molecules across biological ocular membranes. The various transporters present are on various ocular tissues, it is reported by various investigators. However, in the present article, we have focused on the transporters that are localized in the epithelia of the cornea. These ocular transporters may be agreeable to bind and transport specific-targeted ligands attached to drug moieties. Mainly two types of transporter systems are of interest in ocular drug delivery i.e., efflux and influx transporters (Gaudana et al. 2009a, b; Mannermaa et al. 2006; Dey et al. 2003a, b).

Efflux transporter

Efflux transporters belong to the ATP-binding cassette superfamily. This transporter decreases the bioavailability of the drug by effluxing the molecules out of the cell membrane and cytoplasm of the eye. Prominent efflux transporters identified in ocular tissues include P-gp. The other transporter is multidrug resistance protein (MRP), and BCRP. The P-gp efflux transporters have an affinity to efflux lipophilic compounds in normal as well as in cancerous cells, possibly leading to the emergence of drug resistance. Expression and functional activity of P-gp were identified on various ocular cell lines and tissues such as the cornea, conjunctiva as well as a retinal epithelial cell (Kawazu et al. 1999; Dey et al. 2003a, b). The functional activity of MRP transporter are similar to P-gp, but it effluxes only organic anions and conjugated compounds. There are mainly nine isoforms MRP, only three were recognized in ocular tissues i.e., MRP1, MRP2, and MRP5. MRP1 was present in rabbit conjunctival epithelial cells. The efflux pump inhibition can be caused by the following mechanisms (Dey et al. 2004) (Figs. 6.3 and 6.4).

- By blocking of drug binding sites competitively, noncompetitively or allosterically
- By interfering with ATP hydrolysis
- By altering integrity of cell membrane lipids

Influx Transporter

Influx transporters belong to the solute carrier (SLC) superfamily. The amino acid (SLC1, SLC6, and SLC7) and peptides are most commonly pertinent influx

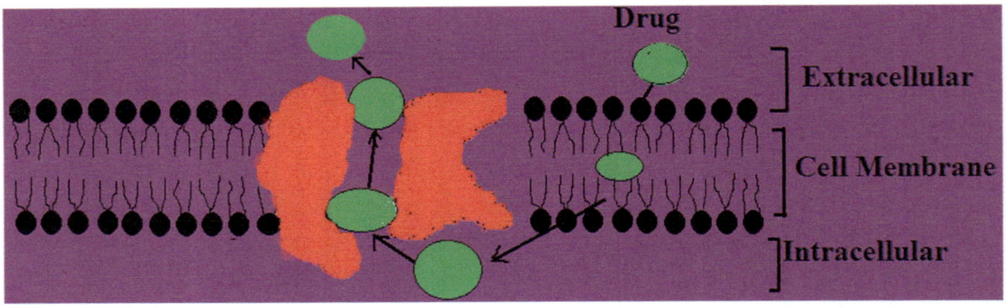

Fig. 6.3 Transportation of drug molecule into the cell

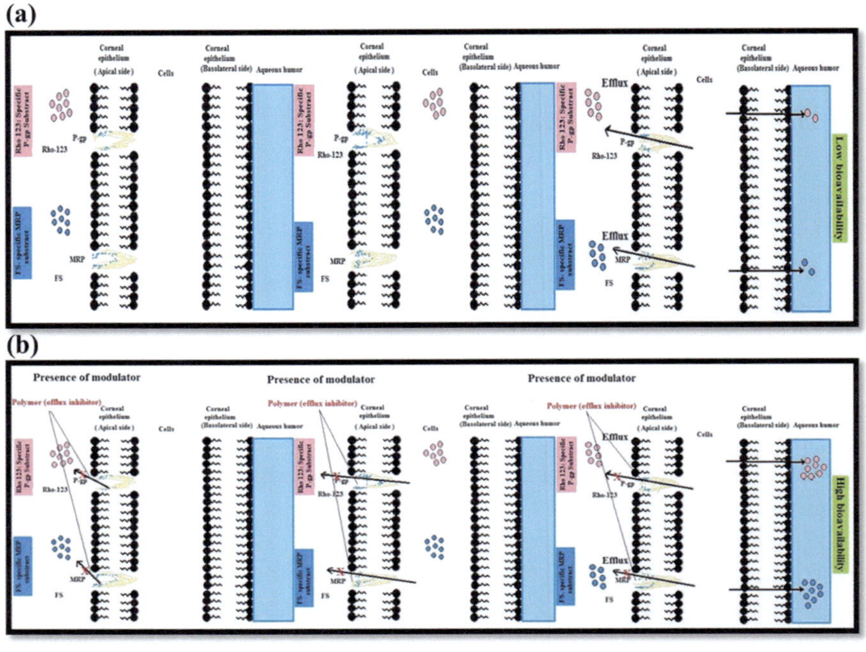

Fig. 6.4 Proposed mechanism of ocular drug transport **a** absence of modulator **b** presence of modulator

transporters for ocular drug delivery. Other types of transporter includes vitamins, glucose, lactate, and nucleoside/nucleobases. These proteins transporter may have a putative role in ocular drug delivery along with their physiological role of transporting various amino acids and nutrients into ocular tissues. Influx transporters assist the transportation of essential nutrients and xenobiotics across biological membranes of the eye.

Nanotechnology Exploited for Transcorneal Drug Delivery

The development of a wide spectrum of nanoscale technologies is beginning to change the scientific landscape in terms of disease diagnosis, treatment, and prevention. These technological innovations, referred to as nanomedicines by the National Institutes of Health, have the potential to turn molecular discoveries arising from genomics and proteomics into a widespread benefit for patients. Nanoparticles can mimic or alter biological processes (e.g., infection, tissue engineering, de novo synthesis, etc.).

The aim of targeted drug delivery and controlled release is to manage better drug pharmacokinetics, pharmacodynamics, non-specific toxicity, immunogenicity and biorecognition of systems in the quest for improved efficacy.

Nanoparticles (NPs), as the very name implies, are particles varying in size from 10 to 1000 nm and depending on the end use, may or may not contain a drug molecule. The drug may be attached to a nanoparticle matrix, or dissolved, encapsulated and entrapped, giving rise to different terminologies as nanoparticles, nanospheres or nanocapsules. All these terms signify their most general characteristic, i.e. they are nanosize particles. Drug loaded nanoparticles (DNPs) constitute an almost versatile drug delivery system, with their ability to overcome physiological barriers and guide the drug to specific cells or intracellular compartments either by passive or ligand-mediated targeting mechanisms (De compose et al. 2001; Quaranta et al. 2013a, b). For ophthalmic applications, properly formulated DNPs are reported to provide ease of application just like eye drop solutions, with the added advantage of being patient friendly, due to less frequent application and extended duration of retention in the extraocular portion. Although the size of the nanoparticles is in the colloidal range which is more precisely accepted to fall between 1 nm and 0.5 μm for ophthalmic formulations, such a preparation may contain larger particles albeit within the colloidal range stated earlier. Terminologies like lyophilic and lyophobic have been used to characterize the dispersion medium, lyophilic systems are usually easier to prepare and have greater stability. Thus, DNPs being organic molecules disperse readily upon addition to the dispersion medium to form colloidal dispersions.

Polymeric Nanoparticles

Polymeric nanoparticles have attracted the interest of many research groups and have been utilized in an increasing number of fields during the last decades. The polymeric nanoparticle enhanced the corneal permeation through enhancing precorneal residence time of the drug. Different polymers can be used to fabricate polymeric NPs such as biodegradable polymers like polylactide (PLAs), poly (D,L-lactides), polylactic-co-glycolic acid, E-caprolactone, (Fessi et al. 1989; Kumari et al. 2010; Addo et al. 2010, 2015; Aksungur et al. 2007; Gupta et al. 2011), polyacrylamide, polycyanoacrylate and polymethylmethacrylate (Zimmer et al. 1991; Wenger et al. 2011) and natural polymers like chitosan (Jain et al. 2013), gelatin (Vandervoort and Ludwig 2004), sodium alginate (Zhu et al. 2012), albumin (Zimmer et al. 1994; Merodio et al. 2002) and tamarind kernel polysaccharide (Kaur et al. 2012) can be used effectively for efficient drug delivery to the ocular tissues. Many researchers conducted the study on nanoparticulate drug delivery system and found that it prevents the degradation of the drug in the ocular environment and release of drug over an extended period of time which gives the desired effect (Gupta et al. 2010; Javadzadeh et al. 2010).

Poly (Alkyl Cyanoacrylate) (PACA) Nanoparticles

Poly (alkyl cyanoacrylate) nanoparticles particles possess properties of biodegradation and bioadhesion, making them of considerable interest as possible drug carriers for controlled ocular drug delivery and drug targeting.

Zimmer et al. (1994) developed pilocarpine nitrate loaded polybutylcyanoacrylate nanoparticles. The pharmacokinetic study in the eyes of New Zealand white rabbits showed that pilocarpine nitrate loaded polybutylcyanoacrylate nanoparticles exhibited an increase of 23 % in pilocarpine levels in aqueous humor and prolonged $t_{1/2}$ compared to the aqueous control solution.

In another study, adsorption of pilocarpine onto polybutylcyanoacrylate nanoparticles enhanced the miotic response by about 22 % compared to the control aqueous drug solution (Harima et al. 1986).

Diepold et al. incorporated pilocarpine into polybutylcyanoacrylate nanoparticles and evaluated the aqueous humor drug levels and the intraocular pressure-lowering effects using three models (the water-loading model, the alpha-chymotrypsin model, and the betamethasone model) in rabbits. Pilocarpine loaded polybutylcyanoacrylate nanoparticles exhibited about 33 % increment of miotic response while the miotic time increased from 180 to 240 min compared to the control solution (Diepold et al. 1989).

Vidmar et al. studied the intraocular pressure-lowering effects of pilocarpine hydrochloride loaded poly (lactic acid) microcapsules. Developed poly (lactic acid) microcapsules of pilocarpine hydrochloride prepared by a solvent precipitation method prolonged miosis about 4 h in comparison to control solution in Female albino rabbits (Vidmar et al. 1985).

A significant improvement in the bioavailability of pilocarpine was attained by co-administering the pilocarpine-loaded albumin nanoparticles with the viscous bioadhesive polymer mucin (Zimmer et al. 1995).

In a clinical study with Piloplex (latex emulsion of pilocarpine hydrochloride), a lower level of the drug with less fluctuation compared to the corresponding control solution was observed on the third day of treatment. This study involving nine subjects showed a reduction by 5.25 mmHg of the average diurnal intraocular pressure (IOP) value compared to the control. Only one out of 30 patients complained of a local sensitivity reaction with Piloplex in the yearlong study (Ticho et al. 1979a, b).

Similar results were obtained in yet another study involving 50 patients, where 67.6 % of the eyes treated with the formulation were under control, while only 45.2 % were under control with the pilocarpine solution (Ticho et al. 1979a, b).

Poly-E-caprolactone nanocapsules also showed good performance in increasing the ocular availability of drugs such as metipranolol (Losa et al. 1993) and betaxolol (Marchal-Haussler et al. 1992) while suppressing their systemic absorption.

Losa et al. developed PECL nanocapsules of metipranolol. The developed nanocapsule exhibited a significant reduction in the intraocular pressure similar to the commercial ophthalmic solution of the drug, but the systemic side effects, studied by evaluation of the cardiovascular effects, were significantly suppressed with the nanocapsules. The heart rate reached normal values within an hour of administration of nanocapsules versus the commercial eye drops, which showed pronounced bradycardia for more than 2 h (Losa et al. 1992).

Marchal-Heussler et al. developed carteolol loaded PECL nanoparticle and nanocapsule (with a TiO_5 oily core) for the treatment of glaucoma. Both formulations demonstrated a pronounced decrease in the intraocular pressure (IOP) compared to the commercial aqueous solution in rabbits with induced intraocular hypertension. The PECL carriers increase the residence time of the drug, enhance the corneal uptake of the drug in unionized form, and decrease the systemic side effects (Marchal-Heussler et al. 1993).

Chitosan Nanoparticles

Kao et al. developed pilocarpine-loaded chitosan/Carbopol nanoparticles. Pilocarpine loaded in nanoparticles showed an initial burst release followed by a continuous and sustained release for 24 h. The in vivo miotic study of developed nanoparticle in New Zealand albino rabbits exhibited the excellent extended miosis effect of pilocarpine (Kao et al. 2006).

Wadhwa et al. developed dorzolamide hydrochloride (DH) and timolol maleate (TM) loaded hyaluronic acid (HA) modified chitosan (CS) nanoparticles (CS-HA-NPs) for the management of glaucoma. Hyaluronic acid provides a synergistic effect for mucoadhesion in association with chitosan. CS-NPs and CS-HA-NPs exhibited a significant reduction in IOP level compared to a plain solution of a drug when administered in glaucomatous male albino rabbits (Wadhwa et al. 2010).

Singh and Shinde developed brimonidine tartrate (BT) loaded chitosan nanoparticles (BT-CS-NPs) for controlled delivery of BT to the ocular membrane. BT-CS-NPs exhibited a significantly higher reduction in IOP compared to marketed formulation (alphagan® P) in glaucomatous female New Zealand rabbits. Draize eye test showed no redness or sign of irritation after instillation of the BT nanoparticles. Furthermore, BT-CS-NPs exhibited initial burst release followed by a prolonged release of BT (Singh and Shinde 2011).

Jain et al., developed betaxolol hydrochloride loaded chitosan nanoparticles (CS-NPs) for ocular delivery and studied their anti-glaucoma efficacy. In vivo pharmacodynamic study in dexamethasone-induced glaucoma model in Male New Zealand albino rabbits showed that betaxolol hydrochloride loaded CS-NPs exhibited gradual reduction of IOP reaching peak value of 9.9 ± 0.5 mm Hg, equivalent to 36.39 ± 1.84 % reduction in IOP compared to control at the end of 5 h which was significant compared to marketed formulation (Jain et al. 2013).

Lin et al. developed pilocarpine-loaded chitosan-PAA nanosuspension using template polymerization of acrylic acid (AA) in a chitosan solution. Both in vitro and in vivo studies in New Zealand albino rabbits revealed that the prepared nanoparticle suspension exhibited better sustained release of pilocarpine than either simulated tear fluid or commercial eye drops (Lin et al. 2007).

Singh et al. prepared pH-triggered polymeric nanoparticulate in situ gel for ophthalmic delivery of acetazolamide by nanoprecipitation method and Carbopol 934P as a gelling agent. The optimized formulation exhibited 3.5-fold (74.50 $\mu g/cm^2$) higher permeation than eye drops (20.08 $\mu g/cm^2$), prolonged precorneal residence time, significant decrease in IOP in comparison to eye drops and sustained drug release along with higher in vitro efficacy, safety and patient compliance (Singh et al. 2014).

Eudragit Nanoparticles

Bhagav et al., developed brimonidine tartrate loaded Eudragit nanoparticles by double emulsion-solvent evaporation technique for the treatment of open-angle glaucoma. The developed nanoparticles were subjected to in vivo intraocular pressure-lowering efficacy studies by administering aqueous dispersion of nanoparticles into the lower cul de sac of glaucomatous rabbits. The selected nanoparticle upon in vivo ocular irritability and tolerability tests were well tolerated with no signs of irritation. The nanoparticle formulations showed a reduction in the elevated IOP in rabbits with glaucoma for a longer period of time. Selected nanoparticles exhibited 7 folds higher AUC (ΔIOP vs. t) value compared to the eye drop preparations (Bhagav et al. 2011).

Poly (DL-Lactide-Co-Glycolide) Nanoparticles

Musumeci et al. developed Melatonin loaded poly (D,L-lactide-co-glycolide) (PLGA) and PLGA-poly(ethylenglycole) (PEG) nanoparticles (NPs) to prolong the pharmacological effects of melatonin to modulate the IOP. The hypotensive effect was evaluated by measuring IOP during 24 h after instillation in the eye of Male albino rabbits of the New Zealand strain in comparison with a melatonin aqueous solution at the same concentration (0.08 %, w/v). Their developed NPs showed good ocular tolerability in rabbit eye using biomicroscopy. Melatonin-loaded PLGA-PEG NPs exhibited 5 mmHg IOP reduction up to 8 h (Musumeci et al. 2013).

Warsi et al. developed dorzolamide (DZ)-loaded PLGA nanoparticle by using two different emulsifying agents (PVA and vitamin E TPGS) for the treatment of glaucoma. Nanoparticles emulsified with vitamin E TPGS (DZ-T-NPs) were found to have higher encapsulation efficiency (59.8 ± 6.1 %) when compared to PVA as emulsifier (DZ-P-NPs). Y-Scintigraphy studies showed the reduced corneal clearance, as well as nasolacrimal drainage in comparison to drug solution. Also, efficacy study revealed that DZ-P-NPs and DZ-T-NPs markedly reduced the intraocular pressure after a single topical instillation into the eye (Warsi et al. 2014) (Table 6.2).

Lipid-Based Nanoparticles

The successful implementation of nanoparticles for drug delivery depends on their ability to penetrate through several anatomical barriers, sustained release of their contents and their stability in the nanometer size. However, the scarcity of safe polymers with regulatory approval and their high cost have limited the widespread application of nanoparticles to clinical medicine.

To overcome these limitations of polymeric nanoparticles, lipids have been put forward as an alternative carrier, particularly for lipophilic pharmaceuticals. These lipid nanoparticles are known as solid lipid nanoparticles (SLNs), nanostructured lipid carriers (NLCs) which are attracting the wide attention of formulators world-wide.

Solid Lipid Nanoparticles

Solid lipid nanoparticles (SLNs) were introduced in 1991, are used as alternative carrier systems to traditional colloidal carriers for treatment of glaucoma, such as emulsions, liposomes and polymeric micro- and nanoparticles (Uner and Yener 2007; Zhang et al. 2010; Hu et al. 2005; Wasutrasawat et al. 2013; Parhi and Suresh 2010). SLNs indicate lipids, which are used in the development of nanoparticles; it is solid at room temperature and body temperature. They are a new generation of submicron-sized lipid emulsions where the liquid lipid (oil) has been substituted by a solid lipid. SLN offer unique properties such as small size, large surface area, high drug loading and the interaction of phases at the interfaces, and are attractive for their potential to improve the performance of pharmaceuticals, nutraceuticals and other materials (Kalam et al. 2010; Seyfoddin et al. 2010).

Table 6.2 Polymeric nanoparticles exploited for ocular delivery in glaucoma

Drug molecules	Polymer used	Formulation	Inferences	References
Betaxolol	Chitosan	Nanoparticles	Marked reduction in IOP	Jain et al. (2013)
Dorzolamide hydrochloride (DH) or timolol maleate	Hyaluronic acid + chitosan	Nanoparticles	Formulation showed significantly decrease in intra ocular pressure (IOP) due to bioadhesive nature of polymer, which enhanced the corneal contact time	Wadhwa et al. (2010)
Dorzolamide HCl and Pramipexole HCl	Chitosan	Nanoparticles	Formulation showed good mucoadhesion and sustained in vitro release	Papadimitriou et al. (2008)
LCS-NP complex	Chitosan	Nanoparticles	Formulation showed no cytotoxicity on conjunctival epithelial cell line IOBA-NHC and strong cellular uptake by corneal epithelium due to bioadhesive nature of polymer	Diebold et al. (2007)
Metipranolol	Chitosan	Nanoparticles	Decreases the systemic side effect because it decreases or inhibits the nasolacrimal drainage	Losa et al. (1993)
Pilocarpine	Albumin	Nanoparticles	Increased the bioavailability of pilocarpine by about 50–90 % (miosis), and decreased IOP 50–70 %, when compared to pilocarpine solution due to high surface of nanoparticles, it enhanced the corneal contact time and decreases nasolacrimal drainage of drug	Zimmer et al. (1994)
Melatonin	PLGA	Nanoparticles	Formulation showed significant decrease in intraocular pressure due to high surface area of particles and minimizes the drainage of drug through tear fluid	Musumeci et al. (2013)
Brimonidine and timolol maleate	PLGA	Hybrid hydrogel Dendrimer	Formulation showed sustained release of drug over a day. Significantly reduced the intraocular pressure	Yang et al. (2012)

(continued)

Table 6.2 (continued)

Drug molecules	Polymer used	Formulation	Inferences	References
Carteolol	Polyalkylcyanoacrylate (PACA)	Nanoparticles and Nanocapsules	Decrease in IOP was much more pronounced with carteolol incorporated into the colloidal carriers than with the commercial eye drops	Heussler et al. (1993)
Metipranolol	PBCA/PECL	Nanocapsules	10 % reduction in IOP in 6 h	Losa et al. (1993)
Acetazolamide	Lipid	Liposomes	The positively charged and neutral liposomes exhibited greater lowering in IOP and a more prolonged effect than the negatively charged ones in rabbit eye	Hathout et al. (2007)
Acetazolamide	Lipid	liposome	Significant and prolonged decrease in IOP compared to the solution of free drug and plain niosomes	Guinedi et al. (2005)
Acetazolamide	Lipid	Liposome	Strong and sustained reduction in IOP in rabbits	Omaima and Ahmed (1997)
Pilocarpine	Lipid	Liposome	Prolonged duration of the miotic effect as compared to aqueous solutions and non-coated vesicles	Durrani et al. (1992)
Penicillin G	Lipid	Liposome	More than four-fold flux increase across isolated rabbit cornea	Kaur et al. (2004)
Timolol maleate	Non-ionic surfactant	Niosome	It showed 1.7 times higher peak concentration of drug in aqueous humor when compared to timolol maleate solution with 2.34 times AUC	Kaur et al. (2010)
Acetazolamide	Span 60	Niosome	Formulation showed higher aqueous humor drug concentration due to enhanced corneal permeation	Aggarwal et al. (2007)

Nanostructured Lipid Carriers

NLCs were introduced to overcome the potential difficulties with SLNs. The goal was to increase the drug loading and prevent drug expulsion. This could be visualized in three ways. In the first model, spatially different lipids (like glycerides) composed of different fatty acids are mixed. The use of spatially different lipids leads to larger distances between the fatty acid chains of the glycerides and general imperfections in the crystal and thus provides more room for accommodation of guest molecules. The highest drug load could be achieved by mixing solid lipids with small amounts of liquid lipids (oils) (Kalam et al. 2010; Wang et al. 2014a, b).

Wang et al. developed and evaluated Methazolamide (MTZ) loaded solid lipid nanoparticles modified with low molecular weight chitosan for the treatment of glaucoma. In vitro release profile of MTZ from CS-SLN-MTZ exhibited prolonged release pattern. CS-SLN-MTZ exhibited excellent permeation in excised New Zealand albino rabbit cornea. In vivo studies revealed that the CS-SLN-MTZ exhibited significantly higher IOP-lowering effect of (245.75 ± 18.31 mmHg × h) compared to both SLN-MTZ (126.74 ± 17.73 mmHg × h) and commercial product Brinzolamide Eye Drops AZOPT® (171.17 ± 16.45 mmHg × h). CS-SLN-MTZ showed no ocular irritancy sign to the Draize method and the histological examination (Wang et al. 2014a, b).

Chen et al. developed methazolamide loaded phosphate (CaP) nanoparticles for local delivery of methazolamide to the eye. In vitro release study of methazolamide loaded CaP-NPs exhibited release as diffusion-controlled from the CaP-NPs over a period of 4 h. In vivo study indicated that the intraocular pressure (IOP)-lowering effect of the inorganic nanoparticle eye drops lasted for 18 h, which was significantly better than the effect of 1 % brinzolamide eye drops (6 h) (Chen et al. 2010).

Tuomela et al. developed brinzolamide (BRA) loaded nanocrystal formulation. The intraocular pressure (IOP) lowering effect was investigated in vivo using a modern rat ocular hypertensive model and elevated IOP reduction was seen in vivo with all the formulations. The developed nanocrystal exhibited significantly decreased intraocular pressure values in rat ocular hypertension model. Furthermore, all formulations showed advantageous dissolution and absorption behavior.

Leonardi et al., developed melatonin (MEL) loaded cationic solid lipid nanoparticles. The ocular hypotensive effect was evaluated by measuring the intraocular pressure (IOP) during 24 h in Male albino rabbits of the New Zealand strain. MEL loaded SLNs exhibited a significant IOP reduction in the rabbit eye. All the formulations tested in vivo demonstrated a good tolerability. The nanocarrier containing stearic acid was the most effective in terms of IOP reduction (maximum IOP reduction: −7 mmHg), and its effect lasted approximately 24 h (Leonardi et al. 2014).

References

Addo RT, Siddig A, Patel NJ, Siwale R, Akande J, Uddin AU, D'Souza MJ (2010) Formulation, characterization, and testing of tetracaine hydrochloride-loaded albumin-chitosan microparticles for ocular drug delivery. J Microencapsul 27(2):95–104

Addo RT, Yeboah KG, Siwale RC, Siddig A, Jones A, Ubale RV, Akande J, Nettey H, Patel NJ, Addo E, D'Souza MJ (2015) Formulation and characterization of atropine sulfate in albumin-chitosan microparticles for in vivo ocular drug delivery. J Pharm Sci 104(5):1677–1690

Aggarwal D, Pal D, Mitra AK, Kaur IP (2007) Study of the extent of ocular absorption of acetazolamide from a developed niosomal formulation, by microdialysis sampling of aqueous humor. Int J Pharm 338(1–2):21–29

Akhter S, Talegaonkar S, Khan ZI, Jain GK, Khar RK, Ahmad FJ (2011) Assessment of ocular pharmacokinetics and safety of ganciclovir loaded nanoformulation. Biomed Nanotechnol 7:144–145

Aksungur P, Demirbilek M, Denkbas EB, Vandervoort J, Ludwig A, Unlu N (2007) Development & characterization of Cyclosporine A loaded nanoparticles for ocular drug delivery: cellular toxicity, uptake, and kinetic studies. J. Control Rel 151:286–294

Ananthula HK, Vaishya RD, Barot M, Mitra AK (2009) Duane's Ophthalmology. In: Tasman W, Jaeger EA (eds) Bioavailability. Lippincott Williams & Wilkins, Philadelphia

Bhagav P, Upadhyay H, Chandran S (2011) Brimonidine tartrate-eudragit long-acting nanoparticles: formulation, optimization, in vitro and in vivo evaluation. AAPS PharmSciTech 12 (4):1087–1101

Bourlais CL, Acar L, Zia H, Sado PA, Neehan T et al (1998) Ophthalmic drug delivery systems-recent advances. Prog Retin Eye Res 17(1):33–58

Casson RJ, Chidlow GW, John PM, Crowston JG, Goldberg I (2012) Definition of glaucoma: Clinical and experimental concepts. Clin Exp Ophthalmol 40(4):341–349

Chen R, Qian Y, Li R, Zhang Q, Liu D, Wang M, Xu Q (2010) Methazolamide calcium phosphate nanoparticles in an ocular delivery system. Yakugaku Zasshi 130(3):419–424

De-Santis LM Jr (1994) Adrenergic receptor-blocking agents. In: Mauger TF, Craig EL (eds) Havener's ocular pharmacology, 6th edn. Mosby-Year Book, St Louis, MO, pp 84–112

Dey S, Anand BS, Patel J, Mitra AK (2003a) Transporters/receptors in the anterior chamber: pathways to explore ocular drug delivery strategies. Expert Opin Biol Ther 3(1):23–44

Dey S, Patel J, Anand BS et al (2003b) Molecular evidence and functional expression of P-glycoprotein (MDR1) in human and rabbit cornea and corneal epithelial cell lines. Invest Ophthalmol Vis Sci 44(7):2909–2918

Dey S, Gunda S, Mitra AK (2004) Pharmacokinetics of erythromycin in rabbit corneas after single-dose infusion: role of P-glycoprotein as a barrier to in vivo ocular drug absorption. J Pharmacol Exp Ther 311(1):246–255

Diepold R, Kreuter J, Himber J, Gurny R, Lee VHL, Robinson JR, Saettone MF, Schnaudigel OE (1989) Comparison of different models for the testing of pilocarpine eyedrops using conventional eyedrops as a novel depot formulation (nanoparticles). Graefe's Arch Clin Exp Ophthalmol 227:188

Diebold Y, Jarrín M, Sáez V, Carvalho EL, Orea M, Calonge M, Seijo B, Alonso MJ (2007) Ocular drug delivery by liposome-chitosan nanoparticle complexes (LCS-NP). Biomaterials 28 (8):1553–1564

Doina G, Hosking SL, Orgu S (2004) Autonomic nervous system, circadian rhythms, and primary open-angle glaucoma. Surv Ophthalmol 49:491–508

Duijm HF, van den Berg TJ, Greve EL (1997) Choroidal haemodynamics in glaucoma. Br J Ophthalmol 81:735–742

Durrani AM, Davies NM, Thomas M, Kellaway IW (1992) Pilocarpine bioavailability from a mucoadhesive liposomal ophthalmic delivery system. Int J Pharm 88(1):409–415

Fessi H, Puisieux F, Devissaguet JP, Ammoury N, Benitas S (1989) Nanocapsule formation by interfacial polymer deposition following solvent displacement. Int J Pharm 55:R1–R4

Gaudana R, Jwala J, Boddu SH et al (2009a) Recent perspectives in ocular drug delivery. Pharm Res 26(5):1197–1216

Gaudana R, Jwala J, Boddu SH, Mitra AK (2009b) Recent perspectives in ocular drug delivery. Pharm Res 26(5):1197–1216

Geroski DH, Edelhauser HF (2001) Transscleral drug delivery for posterior segment disease. Adv Drug Deliv Rev 52(1):37–48

Gherghel D, Hosking SL, Orgu S (2004) Autonomic nervous system, circadian rhythms, and primary open-angle glaucoma. Surv Ophthalmol 49:491–508

Guinedi AS, Mortada ND, Mansour S, Hathout RM (2005) Preparation and evaluation of reverse-phase evaporation and multilamellar niosomes as ophthalmic carriers of acetazolamide. Int J Pharm 306:71–82

Gupta H, Aqil M, Khar RK, Ali A, Bhatnagar A, Mittal G (2011) Biodegradable levofloxacine nanoparticles for sustained ocular drug delivery. J Drug Target 19(6):409–417

Gupta H, Aqil M, Khar RK, Ali A, Bhatnagar A, Mittal G (2010) Sparfloxacin-loaded PLGA nanoparticles for sustained ocular drug delivery. Nanomed Nanotech Boil Med 6:324–333

Hämäläinen KM, Kontturi K, Auriola S, Murtomäki L, Urtti A (1997) Estimation of pore size and pore density of biomembranes from permeability measurements of polyethylene glycols using an effusion-like approach. J Controlled Release 49:97–104

Harima T, Kreuter J, Speiser P, Boye T, Gurny R, Kubis A (1986) Enhancement of miotic response of rabbits with pilocarpine-loaded polybutylcyanoacrylate nanoparticles. Int J Pharm 33:187

Hathout RM, Mansour S, Mortada ND, Guinedi AS (2007) Liposomes as an ocular delivery system for acetazolamide: in vitro and in vivo studies. AAPS Pharm SciTech 8(1):E1–E12

Hornof M, Toropainen E, Urtti A (2005) Cell culture models of the ocular barriers. Eur J Pharm Biopharm 60(2):207–225

Hu FQ, Jiang SP, Du YZ, Yuan H, Ye YQ, Zeng S (2005) Preparation and characterization of stearic acid nanostructured lipid carriers by solvent diffusion method in an aqueous system. Colloids Surf B 45(3–4):167–173

Huang AJW, Tseng SCG, Kenyon KR (1989) Paracellular permeability of corneal and conjunctival epithelia. Invest Ophthalmol Vis Sci 30:684–689

Hyun JJ, Michelle AJ, Carbia BE, Plummer C, Chauhan A (2013) Glaucoma therapy by extended release of timolol from nanoparticle loaded silicone-hydrogel contact lenses. J Control Release 165:82–89

Jain GK, Pathan SA, Akhter S, Jayabalan N, Talagaonkar S, Khar RK, Ahmad FJ (2011) Microscopic and spectroscopic evaluation of novel PLGA-Chitosan nanoplexes as ocular delivery system. Colloids Surf B Biointerfaces 82(2):397–403. doi:10.1016/j.colsurfb.2010.09.010

Jain K, Kumar RS, Sood S, Dhyanandhan G (2013) Betaxolol hydrochloride loaded chitosan nanoparticles for ocular delivery and their anti-glaucoma efficacy. Curr Drug Deliv 10(5):493–499

Jain S, Thompson JR, Foot B, Tatham A, Eke T (2014) Severe intraocular pressure rise following intravitreal triamcinolone: a national survey to estimate incidence and describe case profiles. Eye (Lond). doi:10.1038/eye.2013.306

Jarvinen T, Pate DW, Lain K (2000) Cannabinoids in the treatment of glaucoma. Pharmacol Ther 295:203–220

Javadzadeh Y, Ahadi F, Davaran S, Mohammadi G, Sabzevari A, Adibkia K (2010) Preparation and physicochemical characterization of naproxen-PLGA nanoparticles. Colloids Surf. B Biointerfaces 81:498–502

Kalam MA, Sultana Y, Ali A, Aqil M, Mishra AK, Chuttani K (2010) Preparation, characterization, and evaluation of gatifloxacin loaded solid lipid nanoparticles as colloidal ocular drug delivery system. J Drug Target 18(3):191–204

Kao HJ, Lin HR, Lo YL, Yu SP (2006) Characterization of pilocarpine-loaded chitosan/carbopol nanoparticles. J Pharm Pharmacol 58(2):179–186

Kaur H, Ahuja M, Kumar S, Dilbaghi N (2012) Carboxymethyl tamarind kernel polysaccharide nanoparticles for ophthalmic drug delivery. Int J Biol Macromol 50:833–839

Kaur IP, Aggarwal D, Singh H, Kakkar S (2010) Improved ocular absorption kinetics of timolol maleate loaded into a bioadhesive niosomal delivery system. Graefes Arch Clin Exp Ophthalmol 248(10):1467–1472. doi:10.1007/s00417-010-1383-0

Kaur IP, Garg A, Singla AK, Aggarwal D (2004) Vesicular systems in ocular drug delivery: an overview. Int J Pharm 269:1–14

Kawazu K, Yamada K, Nakamura M, Ota A (1999) Characterization of cyclosporin A transport in cultured rabbit corneal epithelial cells: P-glycoprotein transport activity and binding to cyclophilin. Invest Ophthalmol Vis Sci 40(8):1738–1744

Kaye GI, Sibley RC, Hoefle FB (1973) Recent studies on the nature and function of the corneal endothelial barrier. Exp Eye Res 15:585–613

Kayseri O, Lemke A, Hernandez-Trejo N (2005) The impact of nanobiotechnology on the development of new drug delivery systems. Curr Pharm Biotechnol 6:3–5

Kowing D, Messer D, Slagle S, Wasik A (2010) Programs to optimize adherence in glaucoma. Optometry 81:339–350

Kumari A, Yadav SK, Yadav SC (2010) Biodegradable polymeric nanoparticles based drug delivery systems. Colloids Surf A 72(1):1–18

Langman MJS, Lancashire RJ, Cheng KK, Stewart PM (2005) Systemic hypertension and glaucoma: mechanisms in common and co-occurrence. Br J Ophthalmol 89(8):960–963

Lee VH, Robinson JR (1986) Topical ocular drug delivery: recent developments and future challenges. J Ocul Pharm 2:67–108

Leonardi A, Bucolo C, Drago F, Salomone S, Pignatello R (2014) Cationic solid lipid nanoparticles enhance ocular hypotensive effect of melatonin in rabbit. Int J Pharm 478(1):180–186

Liaw J, Robinson JR (1992) The effect of polyethylene glycol molecular weight on corneal transport and the related influence of penetration enhancers. Int J Pharm 88:125–140

Liaw J, Robinson JR (1993) Ocular penetration enhancers. In: Mitra AK (ed) Ophthalmic drug delivery systems. Marcel Dekker, New York, pp 369–381

Lin HR, Yu SP, Kuo CJ, Kao HJ, Lo YL, Lin YJ (2007) Pilocarpine-loaded chitosan-PAA nanosuspension for ophthalmic delivery. J Biomater Sci Polym Ed 18(2):205–221

Losa C, Alonso MJ, Vila JL, Orallo F, Martinez J, Saavedra JA, Pastor JC (1992) Reduction of cardiovascular side effects associated with ocular administration of metipranolol by inclusion in polymeric nanocapsules. J. Ocul Pharmacol 8:191

Losa C, Marchal-Heussler L, Orallo F, Vila Jato JL, Alonso MJ (1993) Design of new formulations for topical ocular administration: polymeric nanocapsules containing metipranolol. Pharm Res 10:80–87

Machaand SA, Mitra K (2003) Overview of ocular drug delivery. In: Mitra AK (ed) ophthalmic drug delivery systems, vol 130. Marcel-Dekker, New York, pp 1–12

Mannermaa E, Vellonen KS, Urtti A (2006) Drug transport in corneal epithelium and blood-retina barrier: emerging role of transporters in ocular pharmacokinetics. Adv Drug Deliv Rev 58 (11):1136–1163

Marchal-Haussler L, Fessi H, Devissaguet JP, Hoffman M, Maincent P (1992) Colloidal drug delivery systems for the eye. A comparison of the efficacy of three different polymers: polyisobutylcyanoacrylate, polylactic-coglycolic acid, poly-epsilon-caprolactone. Pharm Sci 2:98

Marchal-Heussler L, Sirbat D, Hoffman M, Maincent P (1993) Poly (caprolactone) nanocapsules in carteolol ophthalmic delivery. Pharm Res 10:386

Maurice DM, Mishima S (1984) Ocular pharmacokinetics. In: Sears MC (ed) Handbook of experimental pharmacology, vol. 69. Pharmacology of the Eye. Springer-Verlag, Berlin-Heidelberg, pp 19–116

Merodio M, Irache JM, Valamanesh F, Mirshahi M (2002) Ocular disposition and tolerance of ganciclovir-loaded albumin nanoparticles after intravitreal injection in rats. Biomaterials 23:1587–1594

Michelson G, Langhans MJ, Groh MJ (1996) Perfusion of the juxtapapillary retina and the neuroretinal rim area in primary open angle glaucoma. J Glaucoma 5:91–98

Musumeci T, Bucolo C, Carbone C, Pignatello R, Drago F, Puglisi G (2013) Polymeric nanoparticles augment the ocular hypotensive effect of melatonin in rabbits. Int J Pharm 440(2):135–140

Nathanson JA (1980) Effects of a potent and specific beta 2-adrenoceptor antagonist on intraocular pressure. PNAS, USA 77(12):7420–7424

Newell DG (1986) Monoclonal antibodies directed against the flagella of Campylobacter jejuni: cross-reacting and serotypic specificity and potential use in diagnosis. J Hyg (Lond) 96(3):377–384

Ohtake Y, Tanino T, Kimura I et al (2004) Long-term efficacy and safety of combines topical antiglaucoma therapy-timolol and unoprostone versus betaxolol and unoprostone. Nippon Ganka Gakkai Zasshi. J Jpn Ophthalmol Soc 108:23–28

Omaima NE, Ahmed HH (1997) Preparation and evaluation of acetazolamide liposomes as an ocular delivery system. Int J Pharm 158:121–125

Papadimitriou S, Bikiaris D, Avgoustakis K, Karavas E, Georgarakis M (2008) Chitosan nanoparticles loaded with dorzolamide and pramipexole. Carbohyd Polym 73(1):44–54

Parhi R, Suresh P (2010) Production of solid lipid nanoparticles—drug loading and release mechanism. J Chem Pharm Res 2(1):211–227

Piltz-Seymour JR, Grunwald JE, Hariprasad SM, Dupont J (2001) Optic nerve blood flow is diminished in eyes of primary open-angle glaucoma suspects. Am J Ophthalmol 132:63–69

Quaranta L, Katsanos A, Russo A et al (2013a) 24-hour intraocular pressure and ocular perfusion pressure in glaucoma; major review. Surv Ophthalmol 58:26–40

Quaranta L, Katsanos A, Russo A, Riva I (2013b) 24-hour intraocular pressure and ocular perfusion pressure in glaucoma; major review. Surv Ophthalmol 58:26–40

Quigley HA (1996) Number of people with glaucoma worldwide. J Ophthalmol 80:389–393

Quigley HA, Broman AT (2006) The number of people with glaucoma worldwide in 2010 and 2020. Br J Ophthalmol 90:262–267

Rapoport SI (1976) Blood-brain barrier in physiology and medicine. Raven Press, New York

Rhee DJ (2013) Glaucoma. In: Porter RS, Kaplan JL (eds) The Merck manual home health handbook. Retrieved 12 Dec 2013

Sahoo SK, Dilnawaz F, Krishnakumar S (2008) Nanotechnology in ocular drug delivery. Drug Discov Today 13:144–151

Sahoo SK, Labhasetwar V (2003) Nanotech approaches to drug delivery and imaging. Drug Discov Today 8:1112–1120

Satilmis M, Orgu S, Doubler B et al (2003) Rate of progression of glaucoma correlates with retrobulbar circulation and intraocular pressure. Am J Ophthalmol 135:664–669

Saxena R, Prakash J, Mathur P, Gupta SK (2002) Pharmacotherapy of Glaucoma. Indian J Pharmacol 34:71–85

Schoenwald RD (1990) Ocular drug delivery. Pharmacokinetic considerations. Clin. Pharm. 18:255–269

Seyfoddin A, Shaw J, Al-Kassas R (2010) Solid lipid nanoparticles for ocular drug delivery. Drug Deliv 17(7):467–489

Shields MB (1992) Adrenergic inhibitors. In: Williams MD (ed) Textbook of glaucoma, 3rd edn. Wilkins, Baltimore, pp 480–499

Singh J, Chhabra G, Pathak K (2014) Development of acetazolamide loaded, pH triggered polymeric nanoparticulate in situ gel for sustained ocular delivery: in vitro, ex vivo evaluation and pharmacodynamic study. Drug Dev Ind Pharm 40(9):1223

Singh KH, Shinde UA (2011) Chitosan nanoparticles for controlled delivery of brimonidine tartrate to the ocular membrane. Pharmazie 66(8):594–599

Stroman GA, Stewart WC, Golnik KC (1995) Magnetic resonance imaging in patients with low-tension glaucoma. Arch Ophthalmol 113:168–172

Susanna R, Basseto FL (1992) Hemorrhage of the optic disc and neurosensorial dysacousia. J Glaucoma 1:248–253

Tataru CP, Purcarea VL (2012) Antiglaucoma pharmacotherapy. J Med Life 5:247–251

Ticho U, Blumenthal M, Zonis S, Gal A, Blank I, Mazor ZW (1979a) Piloplex, a new long-acting pilocarpine polymer salt. A long-term study. Br J Opthalmol 63:48

Ticho U, Blumenthal M, Zonis S, Gal A, Blank I, Mazor ZW (1979b) A clinical trial with Piloplex—a new long-acting pilocarpine compound: preliminary report. Ann Ophthalmol 11:555

Uner M, Yener G (2007) Importance of solid lipid nanoparticles (SLN) in various administration routes and future perspectives. Int J Nanomedicine 2(3):289–300

Urtti A, Pipkin JD, Rork G, Sendo T, Finne U, Repta AJ (1990) Controlled drug delivery for esperimental ocular studies with timolol 2. Ocular and systemic ahsorption in rabbits. Int J Pharm 61:241–249

Urtti A, Rouhiainen H, Kaila T, Saano V (1994) Controlled ocular timolol delivery: systemic absorption and intraocular pressure effects in humans. Pharm Res 11(9):1278–1282
Van Buskirk EM, Cioffi GA (1992) Glaucomatous optic neuropathy. Am J Ophthalmol 113:447–452
Vandervoort J, Ludwig A (2004) Preparation and evaluation of drug-loaded gelatin nanoparticles for topical ophthalmic use. Eur. J. Biopharm 57:251–261
Vidmar V, Pepeljnjak S, Jalsˇenjak I (1985) The in vivo evaluation of poly (lactic acid) microcapsules of pilocarpine in hydrochloride. J Microen 2:289
Wadhwa S, Paliwal R, Paliwal SR, Vyas SP (2010) Hyaluronic acid modified chitosan nanoparticles for effective management of glaucoma: development, characterization, and evaluation. J Drug Target 18(4):292–302. doi:10.3109/10611860903450023
Wang F, Chen L, Zhang D, Jiang S, Shi K, Huang Y, Li R, Xu Q (2014a) Methazolamide-loaded solid lipid nanoparticles modified with low-molecular weight chitosan for the treatment of glaucoma: vitro and vivo study. J Drug Target 22(9):849–858
Wang F, Chen L, Jiang S, He J, Zhang X, Peng J, Xu Q, Li R (2014b) Optimization of methazolamide-loaded solid lipid nanoparticles for ophthalmic delivery using Box–Behnken design. J Liposome Res 24(3):171–181
Warsi MH, Anwar M, Garg V, Jain GK, Talegaonkar S, Ahmad FJ, Khar RK (2014) Dorzolamide-loaded PLGA/vitamin E TPGS nanoparticles for glaucoma therapy: pharmacoscintigraphy study and evaluation of extended ocular hypotensive effect in rabbits. Colloids Surf B Biointerfaces 122:423–431
Wasutrasawat P, Al-Obaidi H, Gaisford S, Lawrence MJ, Warisnoicharoen W (2013) Drug solubilisation in lipid nanoparticles containing high melting point triglycerides. Eur J Pharm Biopharm 85(3):365–371
Wenger Y, Schneider RJ, Reddy GR, Kopelman R, Jolliet O, Philbert MA (2011) Tissue distribution and pharmacokinetics of stable polyacrylamide nanoparticles following intravenous injection in the rat. Toxicol Appl Pharmacol 251:181–190
Zhang R, He R, Qian J, Guo J, Xue K, Yuan YF (2010) Treatment of experimental autoimmune uveoretinitis with intravitreal injection of tacrolimus (FK506) encapsulated in liposomes. Invest Ophth Vis Sci 51(7):3575–3582
Zhu X, Su M, Tang S, Wang L, Liang X, Meng F, Hong Y, Xu Z (2012) Synthesis of thiolated chitosan and preparation nanoparticles with sodium alginate for ocular drug delivery. Mol. Vis. 18:1973–1982
Zimmer AK, Chetoni P, Saettone MF, Zerbe H, Kreuter J (1995) Evaluation of pilocarpine-loaded albumin particles as controlled drug delivery systems for the eye. II. Co-administration with bioadhesive and viscous polymers. J Cont Rel 33:31
Zimmer A, Mutschler E, Lambrecht G, Mayer D, Kreuter J (1994) Pharmacokinetic and Pharmacodynamic aspects of an ophthalmic pilocarpine nanoparticle-delivery-system. Pharm Res 11:1435
Zimmer AK, Kreuter J, Robinson JR (1991) Studies on the transport pathway of PBCA nanoparticles in ocular tissues. J Microencapsul 8(4):497–504
Zimmerman TJ (1993) Topical ophthalmic beta blockers: a comparative review. J Ocul Pharmacol 9:373–384

Chapter 7
RETRACTED CHAPTER: In Vitro and in Vivo Evaluation of Ocular Drugs and Delivery Systems

Ruhi V. Ubale and Richard T. Addo

Abstract Animal models like the rabbit have traditionally been used to study the safety and efficacy of ophthalmic products. However, these models are now steadily being replaced by in vitro and ex vivo models that are more convenient and less expensive. The chapter provides an overview of the physicochemical tests used for ocular products as well as the established and recently constructed ocular models and the advantages and limitations associated with these models.

Keywords In vivo · In vitro · Ex vivo · Ocular evaluation

Introduction

For many years, in vivo animal models have been used in ophthalmic studies, mainly to evaluate the level of irritation and toxicity of administered molecules to the ocular cells and tissue. However, in recent years, due to cost, time, and ethical issues associated with animal use (Hornof et al. 2005), researchers have been encouraged to develop alternative techniques for ocular investigations. These techniques include ex vivo models of deceased animal tissue and in vitro human and animal cell culture models. In addition, in vitro evaluation of ocular delivery systems must be done to determine physicochemical properties such as sterility,

The original version of this chapter was revised: The erratum to this chapter is available at https://doi.org/10.1007/978-3-319-47691-9_11

R.V. Ubale · R.T. Addo (✉)
Department of Pharmaceutical Sciences, School of Pharmacy, Union University, Jackson, TN 38305, USA
e-mail: raddo@uu.edu

R.V. Ubale
School of Pharmacy, Lake Erie College of Osteopathic Medicine, Bradenton, FL, USA

R.T. Addo (ed.), *Ocular Drug Delivery: Advances, Challenges and Applications*, DOI 10.1007/978-3-319-47691-9_7 clarity, particle size and morphology, osmolarity, viscosity, stability, dissolution/release, and toxicity.

In Vitro Evaluation

Sterility Testing

Sterility is a basic requirement of all ocular drug delivery systems. Sterility testing involves inoculation of the drug formulation in either of the following microbiological media, thioglycolate medium that supports the growth of aerobic and anaerobic bacteria, and a medium with hydrolysate of casein and soy that supports the growth of aerobic bacteria and fungi.

Two methods, direct inoculation, and membrane filter method, are used for this testing. Direct inoculation involves transferring a suitable amount of the formulation to the medium, followed by incubation, and observation at specified time intervals. The membrane filter method involves filtering the formulation through a 0.45 μm pore, followed by incubation of the filter in a petri dish with nutrient medium.

Clarity

Clarity is usually tested visually against a white and black background with suitable lighting. Transmittance measurements from a UV-Vis spectrophotometer are also used to test clarity.

Particle Size and Morphology

Optical microscopy, light obscuration particle count testing, dynamic imaging analysis, laser diffraction particle analyzers, electron microscopy, dynamic light scattering, Coulter Counter testing, and nanoparticle tracking analysis are used to evaluate particle size and morphology.

Examination of Content of Substance or Preservative

The examination of drug or preservative content in given formulation is labeled with relevant analytical technique, that is, spectrophotometric method or High-Performance Liquid Chromatography (HPLC).

Stability

Stability testing is performed to assess changes in the active ingredient or formulation due to environmental factors such as temperature, humidity, and light, as well as to set the expiration date and recommended storage conditions. General stability requirements for ocular products are similar to those for other pharmaceutical products and are standardized by International Conference on Harmonisation (ICH).

Generally, active ingredients should be stored in conditions that enable assessment of their stability in varying temperatures and humidity conditions. Storage conditions should correlate with warehousing, transport, and later use conditions.

Drug Release Studies

Bottle method, basket method, paddle method, flow-through cell method and diffusion method with Franz cells are used to study drug release from formulations.

In Vitro Evaluation of Specific Delivery Systems

Gel-Forming Ability for in Situ Gels

This is performed to assess the ability of the formulation to form a gel on the eye surface. A sample of the test formulation is added to a vial containing simulated tear fluid and the sol-gel phase transition is visually observed.

Swelling Index for Inserts

Polymer swelling is vital for activating bioadhesive nature of the polymer. Adhesion increases with polymer hydration until the point when excessive hydration leads to loss of adhesion strength. Swelling also affects drug release from a dosage form. Swelling index is determined by weighing a specific number of inserts and placing them in beakers containing simulated tear fluid at a fixed temperature. At specified time intervals, inserts are taken out, dried with filter paper, and weighed again. This is repeated until no further growth in mass is noted. The extent to which the liquid is taken up called the swelling index, is calculated using the formula

$$\text{Swelling Index} = \left[\frac{(\text{Wt} - \text{W0})}{\text{W0}}\right] \times 100$$

where W0 is the initial sample weight and Wt is the sample weight at time t.

Moisture Absorption and Loss for Inserts

Moisture absorption and loss by inserts is evaluated in order to assess physical stability of inserts in varying moisture conditions. A specific number of inserts are placed in a desiccator that is either maintained at a high moisture level or contains anhydrous calcium chloride to measure moisture absorption and loss, respectively. After a specific period, the inserts are taken out and weighed. Moisture absorption or loss is measured using the following the formula

$$\%\ \text{moisture absorption/loss} = \left[\frac{(\text{Final weight} - \text{Initial weight})}{\text{Initial weight}}\right] \times 100$$

Encapsulation Efficiency for Multicompartment Drug Delivery System

Encapsulation efficiency is evaluated by extracting the drug from the delivery system using centrifugation or centrifugal ultrafiltration, followed by measurement of the drug in the supernatant or filtrate using spectrophotometry or HPLC. The following formula is then used for calculation

$$\text{Encapsulation efficiency}\ (\%) = (\text{Wtotal} - \text{Wfree}) \times \frac{100}{\text{Wtotal}}$$

In Vitro Evaluation of Cell Culture Models

For several years, in vivo models were used to evaluate the toxicity of ocular formulations. However, in recent years, researchers have shifted focus to developing in vitro cell culture and ex vivo models to reduce cost and time, and avoid ethical issues associated with animal use.

In vitro models developed from animal and human primary and immortalized cells are important for developing new approaches to overcome ocular barriers. In vitro, cell-based models offer the advantage of being simple, quick to construct, relatively inexpensive, and reproducible while providing a mechanistic understanding of the results. In addition, in vitro models can be used to evaluate a number of combinations of experimental parameters, which is often not achievable with animal models. These models are commonly used for basic science, pharmaceutical research and development, toxicology, and permeability studies.

Corneal Models

The corneal epithelium (CE) is the outermost layer of the eye; it forms the rate-limiting barrier for transcorneal permeation since the uppermost layers of CE cause over 60 % of corneal transepithelial electrical resistance (TEER). Therefore, models of the cornea are very useful in studying drug permeation, absorption, and ocular toxicity. Since the 1960s, many in vitro epithelial models have been established as replacements to Draize test based on primary and immortalized cell lines of human and animal origin (Huhtala et al. 2008; Postnikoff et al. 2014).

Isolated corneal epithelial cells from rabbits are most commonly used to prepare primary cell culture models. Kawazu et al. created a primary rabbit corneal model that resembled an intact cornea and was used for studying permeability of beta-adrenergic antagonist and levofloxacin (Kawazu et al. 1998, 1999a, b), however the TEER and tight barrier were not comparable to in vivo cornea (Klyce 1972; Marshall and Klyce 1983). Chang et al. developed a primary rabbit corneal epithelial membrane with more distinct barrier properties due to air-interface culture conditions (Chang et al. 2000). In addition, immortalized corneal epithelial cells have been developed from rabbits, rats, and hamsters (Araki et al. 1993, 1994; Okamoto et al. 1995; Halenda et al. 1998).

Primary human corneal epithelial (HCE) cells have also been used to evaluate toxicity by preparing sheets to reconstruct ocular surface (Han et al. 2002, Ramaesh and Dhillon 2003). Ban et al. successfully cultured human corneal limbal cells on a human amniotic membrane to develop a stratified epithelium that showed tight barrier junctions after 28 days in culture (Ban et al. 2003). However, the low availability of donor corneas limits the use of HCEs as an alternative in vitro cornea models. To overcome this, efforts to develop immortalized in vitro models from human cells have increased (Kahn et al. 1993; Araki-Sasaki et al. 1995; Offord et al. 1999; Mohan et al. 2003). Some of the commercially available human-derived CE models include Epi-OcularTM (MatTek), HCE (SkinEthic), and Clonetics (Lonza). EpiOcular is a stratified, squamous epithelial model in which a permeable polycarbonate membrane is used to culture human epidermal keratinocytes from neonatal human foreskin (Jung et al. 2011; Stern et al. 1998). Having an air–liquid interface gives these models the advantage of exhibiting similar morphological and growth characteristics to in vivo conditions (Barile 2010). SkinEthic's™ HCE model is also developed from immortalized human CE mucosa cells cultured at the air-liquid interface using a polycarbonate substrate membrane. Lack of stratum cornea makes the model comparable with the corneal mucosa of the human eye (Van Goethem et al. 2006). Both EpiOcular and SkinEthic are replacing the Draize test as eye irritation test for product development (Mohan et al. 2003). Finally, Clonetics, a model of human CE cells cultured on a permeable membrane is used to assess the corneal penetration of ophthalmic drugs and transepithelial permeability studies (Welty 2011).

Conjunctival Models

Conjunctival models are mostly developed from rabbit primary cells as submerged culture models (Saha et al. 1996a, b, 1998; Basu et al. 1998; Yang et al. 2000a; Scholz et al. 2002) as well as the newer air-liquid interface cell culture systems (Yang et al. 2000b; Gukasyan et al. 2002, 2003). The next step in establishing a functional in vitro conjunctival model is to develop a system derived from human conjunctival cells. Primary human conjunctival culture models are already used for conjunctival tissue transplantation (Scuderi et al. 2002; Sangwan et al. 2003). In addition, two immortalized human conjunctival cell lines are being used to develop an in vitro model of human conjunctival epithelium (Gipson et al. 2003; Diebold et al. 2003).

Retinal Models

Retinal Pigment Epithelium

Although studies with primary cell culture models of retinal pigment epithelium derived from frog (Defoe et al. 1994), rat (Chang et al. 1997), bovine (Hartnett et al. 2003), and chick (Peng et al. 2003) are reported in literature, researchers are mainly interested in models derived from human retinal pigment epithelium cells to avoid species-related applicability problems. Despite the limited availability of human donor eye cells, efforts to use primary culture models of human retinal pigment epithelium for studying drug uptake and transport (Lu et al. 1995), protein expression (Ishida et al. 2004), and cytotoxicity (Ho et al. 1997) are on-going.

Davis et al. developed the first immortalized human retinal pigment epithelium cell line from a primary culture. This cell line mimicked metabolic and morphologic properties of human retinal pigment epithelium, however, because of lack of enzymes in vitro conditions, it is mainly used for cytotoxicity studies (Mannerstrom et al. 2002) rather than drug permeation studies (Hornof et al. 2005). Subsequently, Dunn et al. (1996) developed and characterized the second immortalized human retinal pigment epithelium cell line, ARPE-19, which demonstrated barrier properties and expressed retina-specific markers such as RPE65 and CRLABP. Bodnar et al. developed a third immortalized human retinal pigment epithelium cell line, wherein retinal pigment epithelium cells were transfected with vectors encoding for a human telomerase catalytic unit. As a result, cellular life-spans were extended while retaining normal growth characteristics and gene expression patterns (Bodnar et al. 1998; Jiang et al. 1999).

RPE-19 cells have been used for a variety of in vitro studies such as toxicity studies (Yeung et al. 2004), gene delivery (Haeseleer et al. 2001), polarity studies of proteins (Bailey et al. 2004), and as a model for retinal disease (Yeung et al. 2004). However, the long culture duration is the main limitation of RPE-19 cells; it takes about 2 months for cultured RPE-19 cells to become growth quiescent (Hartnett et al. 2003). In addition, the limited availability of human donor eyes makes validation of RPE models difficult. Also, a phenomenon of de-adaptation, change in morphology and functional appearance of cultured cells, is known to occur; this makes it challenging to develop a model that mimics specialized features of the retinal pigment epithelium (Kaida et al. 2000).

Retinal Capillary Endothelium

Several conditions of the eye such as diabetic retinopathy are related to the breakdown of the inner blood-retinal barrier. Therefore, understanding the underlying factors influencing the permeability of the retinal capillary epithelium is essential for discovering new treatment strategies for such diseases. To date, in vitro models of retinal capillary endothelium have been limited to primary isolated bovine retinal capillary endothelial cells (BRCEC) (Gillies et al. 1995) and immortalized rat retinal capillary endothelial cell line (Hosoya et al. 2001). However, a retinal capillary cell culture model with in vivo barrier properties has not yet been developed. A better understanding of the in vivo barrier properties of the retinal capillary endothelium, detailed characterization of cell lines, and eventually co-culture models will be necessary to establish and scientifically validate a more accurate in vitro model of the inner blood-retinal barrier (Hornof et al. 2005).

Models of Ocular Diseases

To date, a number of animal and cell culture models of ocular diseases have been developed that help to investigate the molecular mechanism of ocular diseases and screen ophthalmic drug candidates. One such disease is age-related macular degeneration; because of the complexities associated with this disease and the unique nature of the human eye, animal models cannot replicate all aspects of the disease (Zeiss 2010). Therefore, retinal pigment epithelium cells are used to investigate the physiology and pathology of the disease. In addition, because cell culture models are experimentally controlled systems, their results are more reproducible than those from animal models (Hornof et al. 2005). A standard primary culture of age-related macular degeneration is human fetal retinal pigment epithelium, which closely models the function and metabolic activity of native retinal pigment epithelium (Ablonczy et al. 2011). Other RPE cell models include RPE derived from stem cells and the immortalized ARPE-19 cell line (Dunn et al. 1996).

Diseases of the optic nerve are also very severe and can result in degeneration of retinal ganglion cell, visual field loss, and potentially, blindness. To date, the most useful glaucoma experimental animal models are monkeys, rats, and mice in which argon laser photocoagulation, diode laser photocoagulation, or translimbal laser photocoagulation are used to induce intraocular pressure. Although animal models are essential to improve our knowledge and to better understand the mechanism of each disease, developing an animal model for a disease is complex, challenging, and these animal models still differ widely in their applicability to the human disease (Levkovitch-Verbin 2004). Therefore, there are a range of cell culture models of glaucoma, which include retinal ganglion cells, mixed retinal cells, transformed retinal cells, and neuronal-like cell lines. Once a culture model is established, multiple mechanisms can be used to simulate injury and study the effectiveness of neuroprotective therapies (Levin 2001).

3D Models of the Eye

In vitro ocular cell culture models have been widely used in various fields of research; toxicological screening, permeability, studies of drug uptake and transport, cell physiology, and tissue engineering. They provide useful data that complement findings from in vivo studies and allow significant reduction in the number of animals used. However, there are intrinsic restrictions associated with these models, mainly attributed to the fact that such systems are cell monolayers grown on a two-dimensional (2D) culture scaffold, which does not take into account the response of cells in the 3D curved environment present in the native ocular tissue (Abbott 2003). The in vivo 3D microenvironment can send signals to a cell through cell–cell or cell–extracellular matrix adhesion and mechanical forces. Consequently, these signals will activate a cascade of interactive events, which will in turn influence cell proliferation, differentiation, cellular structure morphology, and apoptosis (Elliott and Yuan 2011). None of the commercially available in vitro ocular models described in this chapter have been cultured on curved scaffolds to mimic growth conditions of corneal and retinal cells in vivo. In addition, ocular in vitro models are limited to regional parts of the eye and no model has yet been developed as an in vitro ocular equivalent as an organ. Only recently, a study by Postnikoff et al. has taken into account curved cell growth conditions, which focused on the creation of a 3D, stratified, curved epithelium. In this study, human papillomavirus-immortalized HCE cells were cultured on a curved Millicell-HA membrane (mixed cellulose esters; Millipore, Billerica, MA). This culture condition led to a stratified, curved, epithelial model suitable for assessment of cytotoxicity and biocompatibility testing of contact lenses (Postnikoff et al. 2014).

In Vivo/Ex Vivo Evaluation

Animal experimentation is critical when developing ocular drugs and delivery systems. Of all the live animals used for testing such as pigs, monkeys, dogs, cats, etc., rabbits are the most commonly used (Wilson et al. 2015; Wilhelmus 2001). In the 1930s, the FDA developed the rabbit in vivo Draize test for evaluating acute ocular toxicity (Draize and Calvery 1944).

Draize test is an international standard assay wherein New Zealand white rabbits are most commonly used as they are readily obtainable, relatively inexpensive, and have a well-described anatomy with large eyes (Wilhelmus 2001). In this protocol, 0.1 mL of the test substance is applied to only one eye of the conscious rabbit, whereas the untreated eye serves as a control. After 72 h exposure of the test substance on the cornea, conjunctiva, and iris, chemicals can be classified on a subjective scoring that ranges from non-irritating to severely irritating (Draize and Calvery 1944). Despite its gold standard status and being the only validated test for evaluating irritation severity in full range, the Draize test has numerous limitations, including its time consuming and subjective nature of assessment, its lack of repeatability and reproducibility (Davila et al. 1998), high dosage of test materials used (Curren and Harbell 2002), variable estimation of results, and overprediction of human response (Jester et al. 2001), which is mainly related to interspecies differences.

In 1980, Griffith et al. developed the low volume eye irritation test (LVET), as an alternative animal method and following a recommendation from the National Research Council (Griffith et al. 1980). In 1977, the National Research Council committee suggested that the Draize test drawbacks might be more of a volume–response correlation rather than a species–response difference between rabbits and humans. LVET is an alteration to Draize testing in which test substances are only applied to the corneal surface of the rabbit's eye and at a lower volume (0.01 mL vs. 0.1 mL). The rationale for reducing the instilled volume is that it is more representative of the lacrimal fluid volume of both the human and the rabbit eye. Therefore, the LVET was described to cause less stress to tested rabbits and also results could better predict human ocular irritation response (Jester et al. 2001). However, results obtained following exposure to severe irritants in LVET were considered to be an underestimation of results in comparison with the Draize data (Gettings et al. 1996a). Therefore, it is debatable whether to accept LVET as a more accurate test as it lacks the element of exaggeration and overprediction of human responses present in Draize testing (Roggeband et al. 2000). This, on the other hand, raises concerns over assuring public safety due to its moderate protocol (Freeberg et al. 1986). As a result, it is still criticized for using animals and it has yet to be accepted as an alternative test by regulatory agencies.

More recently, ocular organotypic models (Table 7.1) have been used to minimize the use of live animals in experimental studies. These isolated ocular systems retain physiological and biochemical functions of the mammalian enucleated eye or cornea (Cooper et al. 2001; Gettings et al. 1996b). Opacity and permeability of the isolated cornea under the effect of a test substance is quantitatively measured using opacitometry and spectrophotometry, respectively. These measurements combined with histological analysis evaluate the extent of damage caused by the test substance and subsequently drive an eye irritation classification for prediction of potential in vivo ocular irritation of a test substance (Barile 2010).

Table 7.1 ex vivo organotypic models used in ocular testing

Name	Testing objective	Validation status	Limitations	References
Bovine corneal opacity and permeability (BCOP)	Ocular sensitivity and corrosion	EVCAM statement of scientific validity for identification of severe irritants and ocular corrosives	Not as sensitive in distinguishing between mild irritants with the standard protocol	(Barile 2010)
Isolated chicken eye (ICE)	Ocular sensitivity and corrosion	EVCAM statement of scientific validity for identification of severe irritants and ocular corrosives	Possible limitation for solids	(Barile 2010)
Isolated rabbit eye (IRE)	Ocular sensitivity and corrosion	Further review is recommended	Possible limitation for solids	(Barile 2010)

Burton et al. developed the first ocular organotypic model known as the isolated rabbit eye (IRE) test method (Burton et al. 1981). The IRE, also known as rabbit enucleated eye test, was originally used to detect irreversible eye damage caused by severe irritants (Burton et al. 1981). IRE protocols have developed over time and have been widely assessed by regulatory bodies (e.g., The European Commission/British Home Office). In 1997, an evaluation of the test concluded that the assay lacks the ability to predict irritation over the full range and can only be useful for evaluating severe irritants (Chamberlain et al. 1997). To date, IRE is mainly used for nonregulatory optimization studies as it is not characterized as a valid assay for ocular irritancy classification (Scott et al. 2010). In response to the deficiencies associated with IRE, Prinsen and Koëter developed the isolated chicken eye (ICE) test method (Prinsen and Koeter 1993). Chicken eyes are readily available from slaughterhouses with consistent quality and dimensions that make them a practical replacement for IRE. Toxic responses are measured by changes in opacity, fluorescein retention, thickness of tissue upon swelling, and assessment of changes related to the surface of the tissue. In addition, in 1992, Gautheron et al. developed the bovine cornea opacity and permeability (BCOP) assay based on methods originally developed by Muir (Muir 1984) and Tchao (Gautheron et al. 1992). The BCOP assay was internationally accepted in 2009 and its scientific suitability is recognized in identifying substances that can cause serious damage as well as substances categorized as nonirritants. Using porcine cornea in BCOP is advantageous as it more accurately resembles human cornea in terms of thickness and structure and has also been frequently used in ocular studies (Van den Berghe et al. 2005). These models have been able to generate promising results with fewer ethical concerns and at reduced costs. However, they all share the mutual drawback that anatomical and physiological differences among interspecies are still associated with these tests. In addition, these models are only limited to evaluating the corneal effects of the substances and not the systemic effects. In 2007, the scientific advisory committee of the European Center for the Validation of Alternative Methods (ECVAM) issued a statement on organotypic ex vivo assays as ocular screening tests to detect possible corrosives and severe irritants. Based on this statement, both BCOP and ICE test methods are scientifically valid to identify severe ocular irritants, whereas validation of the IRE method required additional work to be performed and further review was recommended (SAC). Thuret et al. used an innovative bioreactor for storage of ex vivo cornea, which maintained intraocular pressure and continuously renewed medium. This allowed rapid reduction of stromal swelling and improvement of endothelial cell viability in comparison to the corneal immersion in a sealed flask (G THURET).

Conclusion

To minimize the use of animals, a great amount of research has been dedicated to the development of non-animal alternatives wherever necessary. Alternatives to animal studies are considered as anything from complete to partial replacement of live animals in biomedical research and experimental studies (Dewhurst and Kojic 2011). An apparent 40 % decrease in animal use and a simultaneous increase in the use of tissue culture and biotechnology show that scientifically valid non-animal techniques are implementable (Badyal and Desai 2014).

The development of alternative ex vivo ocular models has made important contributions to biological research. The BCOP and the ICE test methods have been in development since the early 1990s and are the first ex vivo ocular safety test methods that have been validated by the regulators. In both cases, the animal eyes used in both test methods are slaughterhouse waste, therefore animals were not specifically euthanized to obtain these tissues. The use of these two assays alone could reduce the use of live animals for eye safety testing by 10 % or more (Hood 2008). In addition, the use of in vitro platforms has been greatly attributed to obvious cost and ethical advantages over in vivo models. Finally, the development of 3D in vitro culture models that more closely replicates in vivo and complements 2D cell culture and animal model findings, will help researchers to feel confident that final decisions based on in vivo ocular models are well supported.

References

Huhtala ALS, Tähti H, Uusitalo H (2008) Corneal models for the toxicity testing of drugs and drug releasing materials. Devices TMB

Abbott A (2003) Cell culture: biology's new dimension. Nature 424(6951):870–872

Ablonczy Z, Dahrouj M, Tang PH, Liu Y, Sambamurti K, Marmorstein AD et al (2011) Human retinal pigment epithelium cells as functional models for the RPE in vivo. Invest Ophthalmol Vis Sci 52(12):8614–8620

Araki K, Ohashi Y, Sasabe T, Kinoshita S, Hayashi K, Yang XZ et al (1993) Immortalization of rabbit corneal epithelial cells by a recombinant SV40-adenovirus vector. Invest Ophthalmol Vis Sci 34(9):2665–2671

Araki K, Sasabe T, Ohashi Y, Yasuda M, Handa H, Tano Y (1994) Immortalization of rat corneal epithelial cells by SV40-adenovirus recombinant vector. Nippon Ganka Gakkai Zasshi 98 (4):327–333

Araki-Sasaki K, Ohashi Y, Sasabe T, Hayashi K, Watanabe H, Tano Y et al (1995) An SV40-immortalized human corneal epithelial cell line and its characterization. Invest Ophthalmol Vis Sci 36(3):614–621

Badyal DK, Desai C (2014) Animal use in pharmacology education and research: the changing scenario. Indian J Pharmacol 46(3):257–265

Bailey TA, Kanuga N, Romero IA, Greenwood J, Luthert PJ, Cheetham ME (2004) Oxidative stress affects the junctional integrity of retinal pigment epithelial cells. Invest Ophthalmol Vis Sci 45(2):675–684

Ban Y, Cooper LJ, Fullwood NJ, Nakamura T, Tsuzuki M, Koizumi N et al (2003) Comparison of ultrastructure, tight junction-related protein expression and barrier function of human corneal epithelial cells cultivated on amniotic membrane with and without air-lifting. Exp Eye Res 76 (6):735–743

Barile FA (2010) Validating and troubleshooting ocular in vitro toxicology tests. J Pharmacol Toxicol Methods 61(2):136–145

Basu SK, Haworth IS, Bolger MB, Lee VH (1998) Proton-driven dipeptide uptake in primary cultured rabbit conjunctival epithelial cells. Invest Ophthalmol Vis Sci 39(12):2365–2373

Bodnar AG, Ouellette M, Frolkis M, Holt SE, Chiu CP, Morin GB et al (1998) Extension of life-span by introduction of telomerase into normal human cells. Science 279(5349):349–352

Burton AB, York M, Lawrence RS (1981) The in vitro assessment of severe eye irritants. Food Cosmet Toxicol. 19(4):471–480

Chamberlain M, Gad SC, Gautheron P, Prinsen MK (1997) IRAG working group 1. Organotypic models for the assessment/prediction of ocular irritation. Interagency Regulatory Alternatives Group. Food Chem Toxicol 35(1):23–37

Chang CW, Ye L, Defoe DM, Caldwell RB (1997) Serum inhibits tight junction formation in cultured pigment epithelial cells. Invest Ophthalmol Vis Sci 38(6):1082–1093

Chang JE, Basu SK, Lee VH (2000) Air-interface condition promotes the formation of tight corneal epithelial cell layers for drug transport studies. Pharm Res 17(6):670–676

Cooper KJ, Earl LK, Harbell J, Raabe H (2001) Prediction of ocular irritancy of prototype shampoo formulations by the isolated rabbit eye (IRE) test and bovine corneal opacity and permeability (BCOP) assay. Toxicol In Vitro 15(2):95–103

Curren RD, Harbell JW (2002) Ocular safety: a silent (in vitro) success story. Altern Lab Anim 30 (Suppl 2):69–74

Davila JC, Rodriguez RJ, Melchert RB, Acosta D Jr (1998) Predictive value of in vitro model systems in toxicology. Annu Rev Pharmacol Toxicol 38:63–96

Defoe DM, Ahmad A, Chen W, Hughes BA (1994) Membrane polarity of the Na(+)-K+ pump in primary cultures of Xenopus retinal pigment epithelium. Exp Eye Res 59(5):587–596

Dewhurst DG, Kojic ZZ (2011) Replacing animal use in physiology and pharmacology teaching in selected universities in Eastern Europe–charting a way forward. Altern Lab Anim 39(1):15–22

Diebold Y, Calonge M, Enriquez de Salamanca A, Callejo S, Corrales RM, Saez V et al (2003) Characterization of a spontaneously immortalized cell line (IOBA-NHC) from normal human conjunctiva. Invest Ophthalmol Vis Sci 44(10):4263–4274

Draize JHWG, Calvery HO (1944) Methods for the study of irritation and toxicity of substances applied topically to the skin and mucous membranes. J Pharmacol Exp Therapeut 82:377–390

Dunn KC, Aotaki-Keen AE, Putkey FR, Hjelmeland LM (1996) ARPE-19, a human retinal pigment epithelial cell line with differentiated properties. Exp Eye Res 62(2):155–169

Elliott NT, Yuan F (2011) A review of three-dimensional in vitro tissue models for drug discovery and transport studies. J Pharm Sci 100(1):59–74

Freeberg FE, Nixon GA, Reer PJ, Weaver JE, Bruce RD, Griffith JF et al (1986) Human and rabbit eye responses to chemical insult. Fundam Appl Toxicol 7(4):626–634

Thuret GZH, Bernard A, Naigeon N, Nangoum-Fosso T, Nefzaoui C, Piselli S, Acquart S, Garraud O, Campolmi N, Forest F, Dumollard JM, Peoc'h M, Gain P. Revisiting corneal storage using an innovative bioreactor. Acta Ophthalmologica. 91(S252)

Gautheron P, Dukic M, Alix D, Sina JF (1992) Bovine corneal opacity and permeability test: an in vitro assay of ocular irritancy. Fundam Appl Toxicol 18(3):442–449

Gettings SD, Lordo RA, Demetrulias J, Feder PI, Hintze KL (1996a) Comparison of low-volume, Draize and in vitro eye irritation test data. I. Hydroalcoholic formulations. Food Chem Toxicol 34(8):737–749

Gettings SD, Lordo RA, Hintze KL, Bagley DM, Casterton PL, Chudkowski M et al (1996b) The CTFA evaluation of alternatives program: an evaluation of in vitro alternatives to the Draize primary eye irritation test. (Phase III) surfactant-based formulations. Food Chem Toxicol 34(1):79–117

Gillies MC, Su T, Naidoo D (1995) Electrical resistance and macromolecular permeability of retinal capillary endothelial cells in vitro. Curr Eye Res 14(6):435–442

Gipson IK, Spurr-Michaud S, Argueso P, Tisdale A, Ng TF, Russo CL (2003) Mucin gene expression in immortalized human corneal-limbal and conjunctival epithelial cell lines. Invest Ophthalmol Vis Sci 44(6):2496–2506

Griffith JF, Nixon GA, Bruce RD, Reer PJ, Bannan EA (1980) Dose-response studies with chemical irritants in the albino rabbit eye as a basis for selecting optimum testing conditions for predicting hazard to the human eye. Toxicol Appl Pharmacol 55(3):501–513

Gukasyan HJ, Kannan R, Lee VH, Kim KJ (2003) Regulation of L-cystine transport and intracellular GSH level by a nitric oxide donor in primary cultured rabbit conjunctival epithelial cell layers. Invest Ophthalmol Vis Sci 44(3):1202–1210

Gukasyan HJ, Lee VH, Kim KJ, Kannan R (2002) Net glutathione secretion across primary cultured rabbit conjunctival epithelial cell layers. Invest Ophthalmol Vis Sci 43(4):1154–1161

Haeseleer F, Imanishi Y, Saperstein DA, Palczewski K (2001) Gene transfer mediated by recombinant baculovirus into mouse eye. Invest Ophthalmol Vis Sci 42(13):3294–3300

Halenda RM, Grevan VL, Hook RR, Riley LK (1998) An immortalized hamster corneal epithelial cell line for studies of the pathogenesis of Acanthamoeba keratitis. Curr Eye Res 17(3):225–230

Han B, Schwab IR, Madsen TK, Isseroff RR (2002) A fibrin-based bioengineered ocular surface with human corneal epithelial stem cells. Cornea 21(5):505–510

Hartnett ME, Lappas A, Darland D, McColm JR, Lovejoy S, D'Amore PA (2003) Retinal pigment epithelium and endothelial cell interaction causes retinal pigment epithelial barrier dysfunction via a soluble VEGF-dependent mechanism. Exp Eye Res 77(5):593–599

Ho TC, Del Priore LV, Hornbeck R (1997) Effect of mitomycin-C on human retinal pigment epithelium in culture. Curr Eye Res 16(6):572–576

Hood E (2008) Alternative test models: ocular safety assays accepted. Environ Health Perspect 116(9):A381

Hornof M, Toropainen E, Urtti A (2005) Cell culture models of the ocular barriers. Eur J Pharm Biopharm 60(2):207–225

Hosoya K, Tomi M, Ohtsuki S, Takanaga H, Ueda M, Yanai N et al (2001) Conditionally immortalized retinal capillary endothelial cell lines (TR-iBRB) expressing differentiated endothelial cell functions derived from a transgenic rat. Exp Eye Res 72(2):163–172

Ishida BY, Bailey KR, Duncan KG, Chalkley RJ, Burlingame AL, Kane JP et al (2004) Regulated expression of apolipoprotein E by human retinal pigment epithelial cells. J Lipid Res 45(2):263–271

Jester JV, Li L, Molai A, Maurer JK (2001) Extent of initial corneal injury as a basis for alternative eye irritation tests. Toxicol In Vitro 15(2):115–130

Jiang XR, Jimenez G, Chang E, Frolkis M, Kusler B, Sage M et al (1999) Telomerase expression in human somatic cells does not induce changes associated with a transformed phenotype. Nat Genet 21(1):111–114

Jung KM, Lee SH, Ryu YH, Jang WH, Jung HS, Han JH et al (2011) A new 3D reconstituted human corneal epithelium model as an alternative method for the eye irritation test. Toxicol In Vitro 25(1):403–410

Kahn CR, Young E, Lee IH, Rhim JS (1993) Human corneal epithelial primary cultures and cell lines with extended life span: in vitro model for ocular studies. Invest Ophthalmol Vis Sci 34 (12):3429–3441

Kaida M, Cao F, Skumatz CM, Irving PE, Burke JM (2000) Time at confluence for human RPE cells: effects on the adherens junction and in vitro wound closure. Invest Ophthalmol Vis Sci 41(10):3215–3224

Kawazu K, Midori Y, Ota A (1999a) Cultured rabbit corneal epithelium elicits levofloxacin absorption and secretion. J Pharm Pharmacol 51(7):791–796

Kawazu K, Shiono H, Tanioka H, Ota A, Ikuse T, Takashina H et al (1998) Beta adrenergic antagonist permeation across cultured rabbit corneal epithelial cells grown on permeable supports. Curr Eye Res 17(2):125–131

Kawazu K, Yamada K, Nakamura M, Ota A (1999b) Characterization of cyclosporin A transport in cultured rabbit corneal epithelial cells: P-glycoprotein transport activity and binding to cyclophilin. Invest Ophthalmol Vis Sci 40(8):1738–1744

Klyce SD (1972) Electrical profiles in the corneal epithelium. J Physiol 226(2):407–429

Levin LA (2001) Animal and culture models of glaucoma for studying neuroprotection. Eur J Ophthalmol 11(Suppl 2):S23–S29

Levkovitch-Verbin H (2004) Animal models of optic nerve diseases. Eye (Lond) 18(11):1066–1074

Lu SC, Sun WM, Nagineni CN, Hooks JJ, Kannan R (1995) Bidirectional glutathione transport by cultured human retinal pigment epithelial cells. Invest Ophthalmol Vis Sci 36(12):2523–2530

Mannerstrom M, Zorn-Kruppa M, Diehl H, Engelke M, Toimela T, Maenpaa H et al (2002) Evaluation of the cytotoxicity of selected systemic and intravitreally dosed drugs in the cultures of human retinal pigment epithelial cell line and of pig primary retinal pigment epithelial cells. Toxicol In Vitro 16(2):193–200

Marshall WS, Klyce SD (1983) Cellular and paracellular pathway resistances in the “tight” Cl- -secreting epithelium of rabbit cornea. J Membr Biol 73(3):275–282

Mohan RR, Possin DE, Mohan RR, Sinha S, Wilson SE (2003) Development of genetically engineered tet HPV16-E6/E7 transduced human corneal epithelial clones having tight regulation of proliferation and normal differentiation. Exp Eye Res 77(4):395–407

Muir CK (1984) A simple method to assess surfactant-induced bovine corneal opacity in vitro: preliminary findings. Toxicol Lett 22(2):199–203

Offord EA, Sharif NA, Mace K, Tromvoukis Y, Spillare EA, Avanti O et al (1999) Immortalized human corneal epithelial cells for ocular toxicity and inflammation studies. Invest Ophthalmol Vis Sci 40(6):1091–1101

Okamoto S, Ohji M, Kosaku K, Sundarraj N, Hassell JR, Thoft RA et al (1995) Characterization of immortalized rabbit corneal epithelial cells with SV40 large T antigen. Jpn J Ophthalmol 39 (4):323–333

Peng S, Rahner C, Rizzolo LJ (2003) Apical and basal regulation of the permeability of the retinal pigment epithelium. Invest Ophthalmol Vis Sci 44(2):808–817

Postnikoff CK, Pintwala R, Williams S, Wright AM, Hileeto D, Gorbet MB (2014) Development of a curved, stratified, in vitro model to assess ocular biocompatibility. PLoS ONE 9(5):e96448

Prinsen MK, Koeter HB (1993) Justification of the enucleated eye test with eyes of slaughterhouse animals as an alternative to the Draize eye irritation test with rabbits. Food Chem Toxicol 31 (1):69–76

Ramaesh K, Dhillon B (2003) Ex vivo expansion of corneal limbal epithelial/stem cells for corneal surface reconstruction. Eur J Ophthalmol 13(6):515–524

Roggeband R, York M, Pericoi M, Braun W (2000) Eye irritation responses in rabbit and man after single applications of equal volumes of undiluted model liquid detergent products. Food Chem Toxicol 38(8):727–734

Saha P, Kim KJ, Lee VH (1996a) A primary culture model of rabbit conjunctival epithelial cells exhibiting tight barrier properties. Curr Eye Res 15(12):1163–1169

Saha P, Uchiyama T, Kim KJ, Lee VH (1996b) Permeability characteristics of primary cultured rabbit conjunctival epithelial cells to low molecular weight drugs. Curr Eye Res 15(12):1170–1174

Saha P, Yang JJ, Lee VH (1998) Existence of a p-glycoprotein drug efflux pump in cultured rabbit conjunctival epithelial cells. Invest Ophthalmol Vis Sci 39(7):1221–1226

Sangwan VS, Vemuganti GK, Singh S, Balasubramanian D (2003) Successful reconstruction of damaged ocular outer surface in humans using limbal and conjuctival stem cell culture methods. Biosci Rep 23(4):169–174

Scholz M, Lin JE, Lee VH, Keipert S (2002) Pilocarpine permeability across ocular tissues and cell cultures: influence of formulation parameters. J Ocul Pharmacol Ther 18(5):455–468

Scott L, Eskes C, Hoffmann S, Adriaens E, Alepee N, Bufo M et al (2010) A proposed eye irritation testing strategy to reduce and replace in vivo studies using bottom-up and top-down approaches. Toxicol In Vitro 24(1):1–9

Scuderi N, Alfano C, Paolini G, Marchese C, Scuderi G (2002) Transplantation of autologous cultivated conjunctival epithelium for the restoration of defects in the ocular surface. Scand J Plast Reconstr Surg Hand Surg 36(6):340–348

Stern M, Klausner M, Alvarado R, Renskers K, Dickens M (1998) Evaluation of the EpiOcular (TM) Tissue Model as an Alternative to the Draize Eye Irritation Test. Toxicol In Vitro 12 (4):455–461

Van den Berghe C, Guillet MC, Compan D (2005) Performance of porcine corneal opacity and permeability assay to predict eye irritation for water-soluble cosmetic ingredients. Toxicol In Vitro 19(6):823–830

Van Goethem F, Adriaens E, Alepee N, Straube F, De Wever B, Cappadoro M et al (2006) Prevalidation of a new in vitro reconstituted human cornea model to assess the eye irritating potential of chemicals. Toxicol In Vitro 20(1):1–17

Welty AAKWCJSD (2011) Use of Clonetics® Human Corneal Epithelial Cell Model for Evaluating Corneal Penetration and Hydrolysis of Ophthalmic Drug Candidates. Invest Ophthalmol Vis Sci, 52

Wilhelmus KR (2001) The Draize eye test. Surv Ophthalmol 45(6):493–515

Wilson SL, Ahearne M, Hopkinson A (2015) An overview of current techniques for ocular toxicity testing. Toxicology 327:32–46

Yang JJ, Kim KJ, Lee VH (2000a) Role of P-glycoprotein in restricting propranolol transport in cultured rabbit conjunctival epithelial cell layers. Pharm Res 17(5):533–538

Yang JJ, Ueda H, Kim K, Lee VH (2000b) Meeting future challenges in topical ocular drug delivery: development of an air-interfaced primary culture of rabbit conjunctival epithelial cells on a permeable support for drug transport studies. J Control Release. 65(1–2):1–11

Yeung CK, Chan KP, Chan CK, Pang CP, Lam DS (2004) Cytotoxicity of triamcinolone on cultured human retinal pigment epithelial cells: comparison with dexamethasone and hydrocortisone. Jpn J Ophthalmol 48(3):236–242

Zeiss CJ (2010) Animals as models of age-related macular degeneration: an imperfect measure of the truth. Vet Pathol 47(3):396–4131, 2

RETRACTED CHAPTER

Chapter 8
Ocular Delivery of Proteins and Peptides

Lunawati Bennett

Abstract Biotechnology is steadily becoming an important component of the pharmaceutical industry. The numbers of protein and peptide ocular based drugs being developed have increased. The technique to formulate peptide or protein ocular drugs is very different from formulating conventional small molecule drugs. Additionally, the delivery of a peptide or protein-based drugs is particularly challenging due to the issues of protein instability. This chapter discusses the development of proteins and peptides for ocular delivery with an emphasis on their specific considerations and challenges, and a role of specific proteins such as alpha crystallins and thymosin β4 in ocular drug delivery. Possible routes of protein ocular drug delivery are also discussed.

Keywords Ocular peptides delivery · Proteins delivery · Peptides delivery · Route of delivery

Introduction

Proteins are a vital constituent of the body as they perform or are involved in major physiological and biological processes. The revolution of biotechnology has led to the creation of various therapeutic proteins and use of proteins to detect and diagnose diseases. The chemical structure of proteins allows them to perform specific reactions in the body, to increase its efficacy as drugs, and to decrease their adverse effects. However, several disadvantages of employing proteins for

L. Bennett (✉)
School of Pharmacy, Union University, 1050 Union University Drive,
Jackson, TN 38305, USA
e-mail: llbennett@uu.edu

R.T. Addo (ed.), *Ocular Drug Delivery: Advances, Challenges and Applications*,
DOI 10.1007/978-3-319-47691-9_8

therapeutic purposes should also be considered such as: their short half-lives in the blood stream making them necessary to administer repeatedly, their chemical and physical instability, their rapid denaturation in the stomach and intestinal environment, and their short retention in the intended site of action (Mitragotri et al. 2014; Ibraheem et al. 2014).

Therefore, recombinant proteins have been developed and prepared on a large scale to improve protein-based therapy. Synthesizing recombinant proteins could solve many problems. Various techniques have been developed to improve the physicochemical properties of proteins, to protect them from biological degradation, and to deliver them to their target sites effectively. These techniques include: (1) modification of protein structures, (2) inhibit protein degradation by adding enzymatic inhibitors, (3) use of special adjuvants or modification that can enhance protein absorption, and (4) use of different protein encapsulation techniques (Witting et al. 2015). Figure 8.1 shows major hurdles for making proteins drugs.

Due to its unique characteristics, the eye is considered to be an effective protein delivery route. Ocular drug administration protects proteins metabolism by avoiding gastrointestinal and hepatic first pass metabolism which leads to the high bioavailability of proteins and peptides at the intended site of the eye. However, many drawbacks limit the widespread use of ocular protein delivery due to poor eye membrane permeability, hydrophilic property and macromolecular structures of proteins, and the presence of many enzymes in the eye such as protease and amino peptidase that can degrade the proteins and peptides being administered. Several strategies have been used to improve the bioavailability of proteins and peptides as described below (Mitragotri et al. 2014; Vadlapudi et al. 2012).

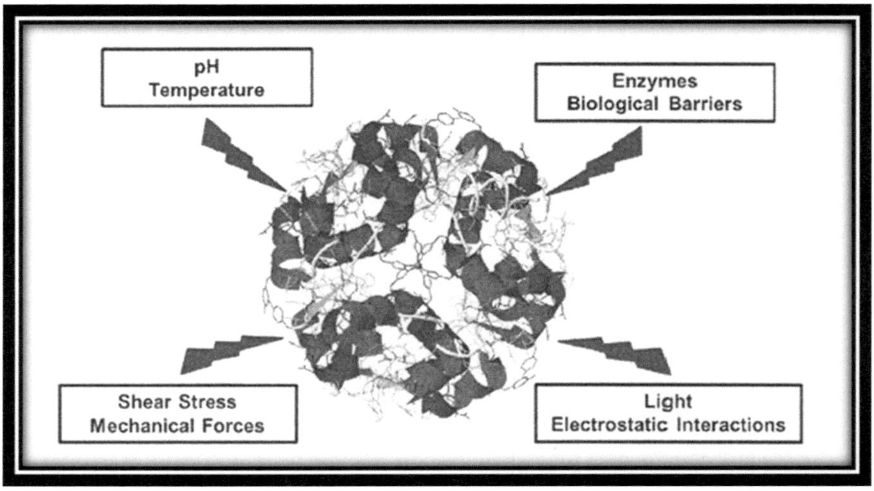

Fig. 8.1 Major hurdles hampering efficient bio macromolecules delivery to target sites occurring during protein formulation, storage and after administration in the human body. With permission from Witting et al (2015)

Use of Penetration Enhancers

Penetration enhancers can be categorized into: (a) chelators (citric acids, salicylates), (b) surfactants (sodium lauryl sulfate, polyoxyethylene-9-lauryl ether), (c) bile salts (sodium glycocholate), (d) fatty acids (oleic acid, capric acid), and (e) non-surfactants (unsaturated cyclic ureas). The mechanism by which penetration enhancer improve bioavailability of peptide drugs include: (1) reducing the barrier functions of mucosal membranes of the eye by changing the structure or properties of these membranes, (2) changing the thermodynamics activity of protein and peptides, and (3) protecting proteins and peptides from proteolytic activity of enzymes present in the eye (Sinha and Trehan 2003).

Use of Enzymatic Inhibitors

Whatever the route used to administer protein or peptide drugs, the enzyme activities are the dominant factor controlling the bioavailability of the given peptides or proteins due to their ubiquitous presence in the body and the sensitivity of proteins or peptides to the enzymatic degradation. Several enzymatic or protease inhibitors such as bacitracin, metalloprotease inhibitors, aspartyl protease inhibitor, cysteine proteinase inhibitor, serine protease inhibitors, and amino peptidase inhibitors have been tested in the ocular products; however, long-term use of these protease inhibitors which include undesirable effects such as causing absorption of unwanted proteins, biodegradation of normal nutritive proteins, and secretion of protease in the body as a result of feedback mechanism should also be considered (Zhou and Po 1991; Ibraheem et al. 2014).

Protein Encapsulation Techniques

Different techniques have been investigated and used for encapsulating therapeutic proteins such as the use of liposomes, double emulsions, polymeric nanoparticles, and solid lipid nanoparticles.

Liposomes are protein carriers that have been studied extensively to deliver suitable vectors due to their biocompatible, biologically inert, minimal immunogenicity, limited toxicity, and small sizes. Liposomes are spherical vesicles consisting of an aqueous core entrapped by one or more lipid bilayers which can be formed by various methods such as thin film hydration, reversed-phase evaporation, detergent dialysis and solvent injection. According to the preparation techniques, liposomes can be multi-, oligo-, or unilamellar with a size varying from several micrometers to 20 nm. Figure 8.2 shows the structure of liposomes (Torchilin and Lukyanov 2003).

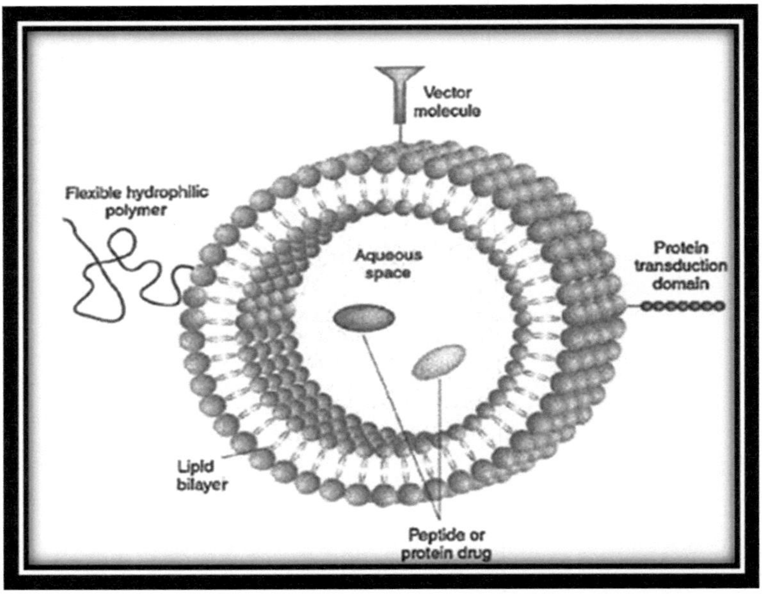

Fig. 8.2 Liposomes grafted with hydrophilic polymers (polyethylene glycol) to increase its circulation half-life. With permission from Torchilin and Lukyanov 2003

Liposomal drug encapsulation provides several advantages for drug delivery to the anterior and posterior segments of the eye because the corneal penetration is increased and longer residence time at the site of action. For example, there was a significant accumulation of liposomal encapsulated vasoactive intestinal peptide (VIP), a potent immunosuppressive, in the posterior segment of the eye 24 h. after injection. The liposomes were internalized by immune cells such as macrophages suggesting that intravitreal injection can increase the ocular immune response. The intravitreal injection of VIP-loaded liposomes resulted in 15 times higher concentrations as compared to the control solution in rat models having ocular inflammation uveitis. To prevent rapid elimination, VIP-loaded liposomes were embedded into hyaluronic acid gel resulting in a long lasting protective effect. Another study reported the use of liposomes with cyclosporine A (CsA), a potent immunosuppressant peptides, which is used to prevent corneal graft rejection, to treat autoimmune uveitis, and to treat dry eye syndrome. CsA loaded liposomes yielded slightly higher drug concentration and longer drug half-life than non-liposomes formulations (Aksungur et al. 2011).

Double emulsions technique involves two stages of an emulsion. In the first stage, the primary emulsion water in oil (W/O) is prepared by adding the aqueous protein solution to the polymer organic solution under high shear conditions (ultra sonication or homogenization). Double emulsion (W/O/W) is produced during the second stage by dispersing again by homogenization the primary emulsion (W/O) in an external aqueous phase containing the chosen stabilizer. Protein loaded particles are obtained after eliminating the organic solvent either by evaporation or by extraction (Ibraheem et al. 2014). Although the double emulsion technique is considered a complex process, it has been widely used to encapsulate proteins in aqueous solution, resulting in high yields and encapsulation efficiencies. Several proteins that have been encapsulated by this method include bovine serum albumin, recombinant human epidermal growth factor (Mprishita and Peppas 2006; Sinha and Trehan 2003).

Polymeric nanoparticles (PNPs) are solid colloidal carriers composed of synthetic, semi-synthetic or natural polymers with size ranging from 10 to 1000 nm. These carriers are usually nanospheres or nanocapsules. In nanospheres the drug is dispersed in a polymeric matrix, whereas nanocapsules are reservoir system in which drugs are confined within a polymeric shell. Proteins can be encapsulated, adsorbed or chemically linked to the surface of PNPs. The physiological and anatomical barriers and enzymatic degradation within the ocular environment limit the efficacy of proteins and peptides administered by ocular route. PNPs have been successfully designed to improve localized ocular delivery by providing sustained release, protection from enzymatic degradation and enhancing precorneal residence time of PNP by adding mucoadhesive such as polyethylene glycol (PEG), carbopol or hyaluronic acid. However, researchers still need to make ideal PNPs that are capable of generating high loading and entrapment efficiency, able to protect protein integrity until it reaches the target site, and able to release the encapsulated protein or peptide in a sustained manner to avoid frequent administrations (Gaudana et al. 2009; Kompella et al. 2013).

Solid lipid nanoparticles (SLNs) combine the advantages of other carrier systems (PNPs, nanoparticles, fat emulsions, and liposomes). These non-toxic drug carriers can efficiently control the release of the drug; both hydrophilic and hydrophobic drugs can be incorporated in SLNs, and can be produced in large scales (Yang et al. 1999). The SLN can be used as parenteral and non-parenteral. In a recent study, coating CsA with SLN cationic polymers such as Eudragit® produced drugs with increased precorneal time due to an interaction between the cationic and anionic mucins present in the mucus layer of the eye surface. Tear kinetic parameters such as cellular uptake, tear film concentration of the drug, the area under the curve (AUC) were significantly higher for the formulation of combo with Eudragit® (cationic NP) and Carbapol® (adhesive) than from known formulation of CsA (Restasis®) (Aksungur et al. 2011).

Alpha-Crystallins

Crystallins constitute a diverse group of proteins that are highly concentrated in differentiated lens fiber cells and augment the refractive power of the transparent lens tissues. In the lens, they are usually found as large aggregates consisting of two types of subunits, alpha (α) and alphaB (αβ) crystallins (Bloemendal and deJong 1991). Recent studies showed α and αβ crystalline mRNAs and proteins are increased in acute induce retinal degeneration. Proposed mechanisms of the roles of α and αβ crystallins include: (1) their over expression is a protective mechanism to prevent retinal cell death during inflammation and (2) preventing ultraviolet light (UVA) induced photoreceptor apoptosis in the retina. However, the up-regulated α and αβ crystallins also indicate neurodegeneration since the presence of antibodies against α and αβ crystalline found in the serum of patients with uveitis and other non-ocular diseases. Intravitreal injection of α crystalline has a positive outcome on retinal cell survival post-trauma such as in optic nerve crush. Alpha crystalline administration prior to or at the time of injury prevented retinal ganglion cell axon degeneration and decreased microglial activation. Figure 8.3. Shows model of alpha-crystallin function during chronic (protective) and acute (impaired) stress conditions (Fort and Lampi 2011; Fort and Freeman 2009;

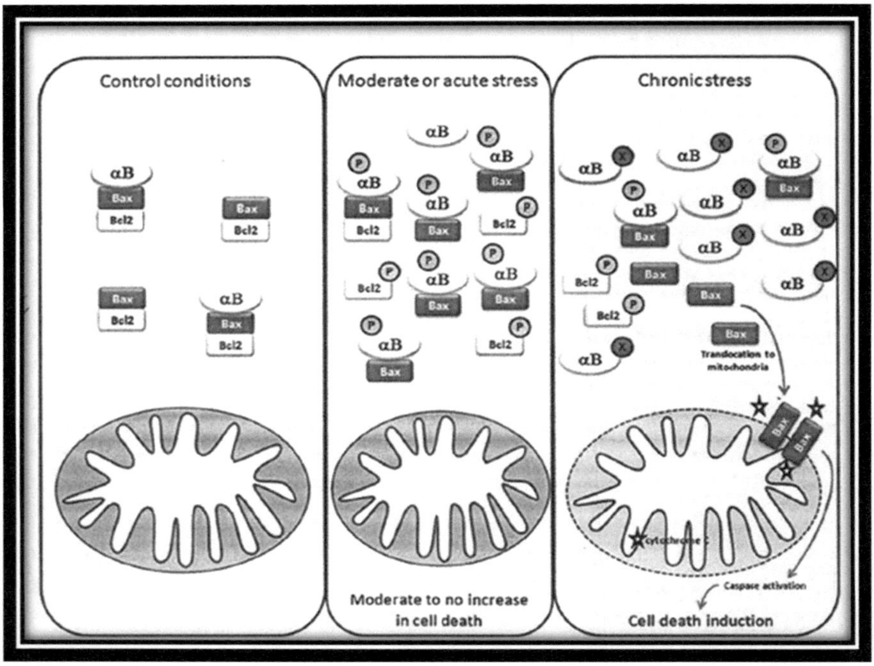

Fig. 8.3 Model of alpha-crystallin function during chronic (protective) and acute (impaired) stress conditions. In a control condition, anti-apoptotic proteins such as bcl-2 interact and prevent the activation of pro-apoptotic proteins such as Bax. When cells are exposed to a moderate or acute stress, different signaling pathways can be activated leading to the inactivation of the anti-apoptotic member of that complex. Parallel pathways lead to increase alphaB crystalline and alpha expression as well anti-apoptotic activity (including P = phosphorylation) through interaction and inhibition of pro-apoptotic proteins such as Bax and Bcl-X. In chronic stress condition such as diabetes retinopathy, retinopathy of prematurity or uveitis, alpha-crystallins undergo post-translational modifications such as nitration, glycation, deamination which inhibits their anti-apoptotic function and lead to cell death induction. With permission from Forth and Lampi (2011)

Table 8.1 Summary of recent discoveries on the role of crystallins in retinal diseases (Fort and Lampi 2011)

Disease	Crystallins	Changes	Associated proteins, effects or localization
Diabetes retinopathy	αA, αβ, β, γ crystallins	Upregulation	Bcl-Xs, Bax, caspases
Retinopathy of prematurity	αβ crystallins	Upregulation	Hif1α, VEGF
Optic nerve injury	αA, αβ crystallins	Downregulation (protective when overexpressed)	Expressed in GC
Autoimmune uveitis	αA, β crystallins	Upregulation	Interaction with nitrated Cyto C
Endophthalmitis	αβ crystallins	Increased cleaved form Increased caspase-3	
Ischemia	αβ crystallins	Upregulation	In GC with heat shock proteins
Age-related macular degeneration	βb2 crystallins	Accumulation in drusen	Lots of protein in drusen
Chemically induced hypoxia	αA, αβ crystallins	Dose responsive (Low dose = upregulate) (High dose = downregulate)	Mitochondrial localization

Santhoshkumar and Murugesan 2009). Table 8.1 summarized recent discoveries on the role of crystallins in retinal diseases.

Thymosin β4

Thymosin β4 is a ubiquitous polypeptide found in all cell types except red blood cells. Thymosin β4 is a multifunctional protein that promotes cell migration, stem cell recruitment and differentiation, protease production, and expression of various regulatory genes such as vascular endothelial growth factor (VEGF), matrix metalloproteinase, hepatocytes growth factor, and anti oxidative enzymes. It inhibits inflammation, microbial growth, scar formation and protects cells from cytotoxic damage. The efficacy of thymosin β4 has been shown in several clinical trials in the treatment of neurotrophic keratitis wounds, to reduce wound size and to decrease symptoms of ocular discomfort in patients with dry eyes (Sosne and Kleinman 2015; Goldstein et al. 2012).

Dry eyes disorders are becoming more common due to many causes such as aging population, increased pollution and post refractive surgery. Current treatment of dry eye is artificial tears, gels, lubricants, tear duct plugs, and anti-inflammatory drugs such as steroids, doxycycline or CsA. Despite these therapies, successful resolution of the problem is limited because none of this treatment fully addresses

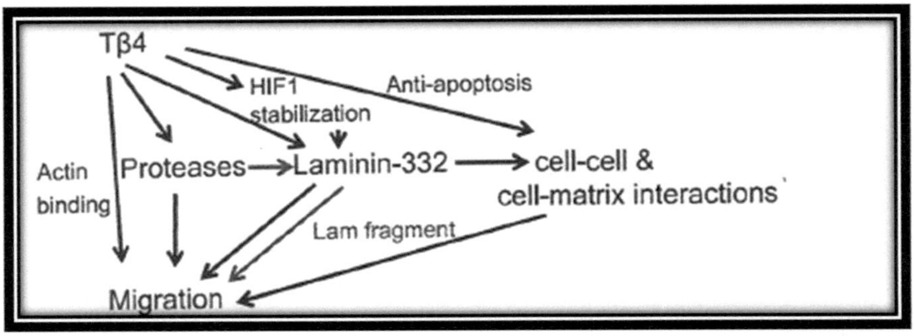

Fig. 8.4 Schematic of how thymosin β4 promotes migration via multiple pathways. Direct migration involves the ability of thymosin β4 to bind actin. Proteases promote migration by releasing chemotactic matrix factors and degrading adhesion receptors. Thymosin β4 induces the synthesis of laminin-332 which is important adhesion and migration factor. One mechanism involves stabilization of the transcription factor HIF1 that binds to the promoter of the laminin-332 chains. Proteases also degrade laminin-332 generating a smaller chemotactic factor. Laminin-332 also stabilizes cell-cell and cell-matrix interactions, which are important for sheet migration over the wound site. The anti-apoptotic activity of thymosin β4 also helps the epithelium to retain its intact structure for sheet migration. With permission from Sosne and Kleinman (2015)

the underlying cause of dry eye which is to promote complete ocular surface repair. Thymosin β4 proposed the mechanism of action to heal dry eye is through its effect in promoting cell migration through actin polymerization, matrix metalloproteinase, synthesis of fibronectin and laminin-332 (Sosne et al. 2006, 2015). Figure 8.4 shows proposed mechanism of how Thymosin β4 promotes migration to repair dry eye diseases.

Routes of Protein Delivery

Systemic Administration

The short-term safety of systemic bevacizumab, a humanized monoclonal antibody that block VEGF-A and its effect on visual acuity and patients with neovascular age-related macular degeneration (AMD) in several clinical trials. Intravenous injection of bevacizumab followed by one or two additional doses given at 2 weeks interval significantly reduced macular edema in 18 AMD patients and improved vision in choroidal neovascularization (CNV) patients. However, repeated systemic bevacizumab may cause adverse effects such as hypertension, neuropathy, congestive heart failure, hemorrhage, neutropenia complications, proteinuria, and arterial thromboembolism. In those experiments, the drug concentration at the vitreous levels was not measured. Further research is needed to measure the vitreous levels of the drugs to determine the right concentration of the drugs to act directly on the choroidal vessels without penetrating into the eye or altering other membrane structures of the eye (Do et al. 2009; Sanharawi et al. 2010).

Recently, systemic drugs such as Remicade®, Enbrel®, Humira® that have been successfully used to treat other conditions such as rheumatoid arthritis, inflammatory bowel disease are have also been tried in treating ocular inflammation such as uveitis, choroidal neovascularization.

Topical Instillation

In adult rats, after topical instillation, antibody fragments of nerve growth factor (NGF) were found in retina and optic nerve. A single chain antibody fragments formulated with viscosity and penetration enhancers was detected in the vitreous at 4 and 12 h but cleared at 12 h following the final eye drop. Other studies used anti-tumor necrosis factor (TNF) single chain antibody showed the drug can reach the retina of a rabbit eye. Several of the recent studies show that topical delivery of protein or peptides with a molecular weight below 30 kDa can reach the vitreous and the retina/choroid area. However, highly concentrated protein solutions have to be instilled due to instability problems of proteins or peptides. Additionally, clinical studies are needed to see if the instilled protein compounds can reach the posterior segment and macula which are difficult to achieve (Sanharawi et al. 2010).

Osmotic Pumps

The effectiveness of osmotic pumps to deliver IgG across the sclera of pigmented rabbits showed that the levels of the IgG in the retina and choroid were significantly higher than the baseline measured over a 28 day period. The elimination of IgG from the choroid and retina followed first order kinetics. The osmotic pump method was also used to deliver an anti-intercellular adhesion molecule 1 (ICAM-1) monoclonal antibody to the retina and choroid with ocular concentration was detected as 700 times higher than plasma concentration. However, the osmotic pumps techniques are invasive and this method of delivering ocular protein has not been used in clinical practice (Ambati et al. 2000).

Subconjunctival Injections

Subconjunctival injections may result in various intraocular drug levels depending on the injected volumes, the site of injection and the animal model used. In a rabbit model, a subconjunctival injection of tissue plasminogen activator (tPA) to treat proliferative vitreous retinopathy or peripheral diabetic retinopathy showed a modification of ruptured ocular barriers. Retrobulbar and sub-tenon injections of nerve growth factor (NGF) to a rat model of retinal degeneration showed a significant delay in retinal degeneration and proliferation of choroid and retina after the treatment (Amaral et al. 2005).

Intravitreal Injections

Intravitreous injections have been the route of choice to administer therapeutic proteins. Anti-VEGF therapies are injected in the vitreous repeatedly to reduce CNV. Several new therapeutic proteins such as neurotropic factors and cytokines to protect retinal cells and to decrease inflammatory process, and anti-angiogenic proteins to inhibit ocular neovascularization and macular edema have been successfully done in clinical trials. In these trials, fibroblast growth factor (FGF-1) injected in the vitreous, or human ciliary neurotropic factor (CNTF) analog can delay photoreceptor loss in retinal degeneration disease, and repeated injection of this protein have a protective effect on the retina (Sanharawi et al. 2010).

A wide variety of the angiogenesis inhibitors: pegaptanib (Macugen®), aflibercept (Eylea®), ranibizumab (Lucentis®), bevacizumab (Avastin®) have been used for the treatment of AMD and macular edema due to central retinal vein occlusion (CRVO). Macugen® and Lucentis® have been approved by the FDA for the treatment of neovascular AMD, but Avastin® is used off-label intravitreally to treat macular edema secondary to neovascular AMD. Another anti-VEGF agent, the VEGF-Trap, has been intensively studied for the management of neovascular edema, and diabetic macular edema The VEGF-Trap is a chimeric VEGFR1-VEGFR2 extracellular portion/Fc fragment fusion protein with higher affinity for VEGF than other anti-VEGF agent. This agent has activity over 10-12 weeks in comparison to Lucentis® activity of 30 days. Currently, several phases 2 and 3 trials are under investigation to evaluate VEGF-Trap eye intravitreal injection in patients with diabetic macular degeneration, neovascular AMD, macular degeneration, and retinal vein occlusion. Table 8.2 showed potential peptide or protein based ocular agents (Nguyen et al. 2009).

Table 8.2 List of potential therapeutic agents for ocular indications (Sanhawarawi et al 2010)

Class	Therapeutic agents	MW (kDa)	Potential indications
Neurotropic agents			RGC, PR degeneration
Growth factors	Nerve growth factor (NGF)	26	Glaucoma, wound healing
	Acid fibroblast growth factor (aFGF, FGF-1)	17	Illumination
	Basic fibroblast growth factor (bFGF, FGF-1)	17–23	Illumination

(continued)

Table 8.2 (continued)

Class	Therapeutic agents	MW (kDa)	Potential indications
Neurotropic factors	Ciliary neurotropic factor (CNTF)	23	Glaucoma, RP
	Brain-derived neurotrophic factor (BDNF)	28–37	Illumination
	Glial cell line neurotropic factor (GDNF)	24	Glaucoma, RP
Anti-inflammatory factors			
Anti-TNF agents			
Anti-TNF antibodies	ESBA 105: anti-TNF single chain Mab	26	Inflammation
	Infliximab: anti-TNF chimeric Mab	165	Uveitis, CNV
	Adalimumab: anti-TNF rhuMab	148	Uveitis
Recombinant TNFR	Etanercept: TNFR-II/Fc fusion protein	130	Uveitis
Immuno regulators			
Cytokines	Interferon gamma	16	Viral infection, uveitis
Peptides	Vasoactive intestinal peptide (VIP)	3.3	Uveitis
Anti-angiogenic factors			
Anti-VEGF antibodies	Bevacizumab: anti-VEGF rhuMab	150	CNV, RVO, CME, PDR
	Ranibizumab: rhuMab VEGF Fab fragments	48	CNV
Recombinant VEGFR	sFlt-1: soluble FMs like tyrosine kinase-1	110	CNV, PDR
Protein	Pigment epithelium-derived factor (PEDF)	50	CNV, ischemia
Antibody	Anti-HER2 rhuMab	148	NV
Anti proliferative	Rituximab:anti CD20 chimeric Mab	145	PIOL
Others			
Fibrinolysing protein	Tissue plasminogen activator (tPA)	69	PVR PDR, endophthalmatis
Model protein	Ovalbumin	45	PVR, PDR, endophthalmatis

Abbreviation: *AMD* Age related macular degeneration; *CME* Cystoid macular edema; *CNV* Chroidal neovascularization; *HER2* Human epidermal growth factor receptor 2; *Mab* Monoclonal antibodies; *MW* Molecular weight; *NV* Neovascularization; *rhuMab* recombinant humanized monoclonal antibodies; *TNF* Tumor necrosis factor alpha; *PIOL* Primary intraocular lymphoma; *PR* Photoreceptors; *PVR* Proliferative vitreoretinopathy; *PDR* Proliferative diabetic retinopathy; *RP* Retinitis pigmentosa; *RGC* Retinal ganglion cells; *RVO* Retinal vein occlusion; *VEGF* Vascular endothelial growth factor

Summary

Proteins and peptides have been studied extensively for the treatment of various ocular chronic diseases due to their high potency and specificity. However, physiochemical properties or proteins and peptides and complex physiology of the non-invasive routes pose significant challenges for the site-specific delivery of these macromolecules to the intended site of the eye.

Polymeric nanoparticles system can promote better absorption of protein via non-invasive routes; however, ideal PNPs remain an active field of research. Surface conjugation of polymeric NPs with antibodies, mucoadhesive polymers, cell penetrating peptides or protease inhibitors may significantly enhance the efficacy of PNPs and protein/peptide-based ocular drug delivery.

References

Aksungur P, Demirbilek M, Denkbas EB et al (2011) Development and characterization of cyclosporine a loaded nanoparticles for ocular drug delivery: cellular toxicity, uptake, and kinetic studies. J Control Rel. 151(3):286–294

Amaral J, Fariss RN, Campos MM et al (2005) Transscleral-RPE permeability of PEDf and ovalbumin proteins: implications for subconjunctival protein delivery. Invest Opthalmol Vis Sci. 46(12):4383–4392

Ambati J, Gragoudas ES, Miller JW et al (2000) Transcleral delivery of bioactive protein to the choroid and retina. Invest Opthamol Vis Sci. 41(5):1186–1191

Bloemendal H, deJong WW (1991) Lens proteins and their genes. Prog Nucleic Acid Res Mol Biol 41:259–281

Do DV, Nguyen QD, Shah SM et al (2009) An exploratory study of the safety, tolerability, and bioactivity of a single intravitreal injection of vascular endothelial growth factor trap-eye in patients with diabetic macular edema. Br J Opthamol. 93(2):144–149

Fort PE, Freeman WM (2009) The retinal proteome in experimental diabetic retinopathy: up-regulation of crystallins and reversal by systemic and periocular insulin. Mol Cell Proteomics 8(4):767–779

Fort PE, Lampi KJ (2011) New focus on alpha-crystallins in retinal neurodegenerative diseases. Exp Eye Res 92:98–103

Gaudana R, Jwala J, Boddu SHS, Mitra AK (2009) Recent perspective in ocular drug delivery. Pharm Res 26(5):1197–1216

Goldstein AL, Hannappel E, Sosne G et al (2012) Thymosin beta 4: a multi-functional regenerative peptide. Expert Opin Biol. 12:37–51

Ibraheem D, Elaissari A, Fessi H (2014) Administration strategies for proteins and peptides. Int J Pharm 477:578–589

Kompella UB, Amrite AC, Ravi RP, Durazo SA (2013) Nanomedicines for back of the eye drug delivery, gene delivery, and imaging. Prog Retin Eye Res. 36:172–198

Mitragotri S, Burke P, Langer R (2014) Overcoming the challenges in administering biopharmaceuticals: formulation and delivery strategies. Nat Rev Drug Discov. 13:655–672

Mprishita M, Peppas NA (2006) Is the oral route possible for peptide and protein drug delivery? Drug Discov Today 11:905–910

Nguyen QD, Shah SM, Browning DJ et al (2009) A phase 1 study of intravitreal vascular endothelial growth factor trap-eye in patients with neovascular age-related macular degeneration. Opthamol. 116(11):2141–2148

Santhoshkumar P, Murugesan R (2009) Deletion of (54) FRAPSWF (61) residues decreases the oligomeric size and enhances the chaperone function of alphaB-crystallin. Biochemistry 48 (23):5066–5073

Sanharawi ME, Kowalczuk L, Touchard E et al (2010) Protein delivery for retinal diseases: from basic considerations to clinical applications. Progr Retin Eye Res 229:443–465

Sinha VR, Trehan A (2003) Biodegradable microspheres for protein delivery. J Control Rel 90:261–280

Sosne G, Albeiruti A, Hollis B et al (2006) Thymosin beta 4 inhibits benzalkonium chloride-mediated apoptosis in corneal and conjunctival epithelial cells in vitro. Exp Eye Res 83:502–507

Sosne G, Kleinman HK (2015) Primary mechanisms of thymosin β4 repair activity in dry eye disorders and other tissue injuries. Invest Opthalmol Vis Sci 56:5110–5117

Sosne G, Kim C, Dunn SP (2015) Thymosin beta 4 significantly improves signs and symptoms of severe dry eye in phase 2 clinical trial. Cornea 34:491–496

Torchilin VP, Lukyanov AN (2003) Peptide and protein drug delivery to and into tumors: challenges and solutions. Drug Discov Today 8:259–266

Vadlapudi AD, Patel A, Cholkar K (2012) Mitra AK. Recent patents on emerging therapeutics for the treatment of glaucoma, age-related macular degeneration, and uveitis. Recent Pat Biomed Eng 5(1):83–101

Witting M, Obst K, Friess W, Hedtrich S (2015) Recent advances in topical delivery of proteins and peptides mediated by soft matter nanocarriers. J Biotech Adv. 33:1355–1369

Yang SC, Lu LF, Cai Y et al (1999) Body distribution in mice of intravenously injected camptothecin solid lipid nanoparticles and targeting effect on brain. J Control Rel 59:299–307

Zhou XH, Po LW (1991) Peptide and protein drugs: therapeutic applications, absorption, and parenteral administration. Int J Pharm 75:97–115

Chapter 9
Progress of Controlled Drug Delivery Systems in Topical Ophthalmology: Focus on Nano and Micro Drug Carriers

Ameeduzzafar Zafar, Javed Ahmad, Richard T. Addo and Sohail Akhter

Abstract Corneal and conjunctival epithelia, along with the tear film, serve as biological barriers to protect the eye from the entrance of potentially harmful substances. These barriers create constraint effective drug medication of ocular diseases with topical ocular formulations. Designing an effective therapy for ocular diseases, especially for the anterior segment, has been considered a challenging task. Nano-drug carriers giving us an array of hope for ocular drug therapy, owing to its potential to improve the ocular retention, controlled release, trans-corneal permeation and thus intra-ocular drug availability. Nanotechnology-based formulation design is important in ocular pharmaceuticals yet knowledge of anatomy and physiology of eyes are critical along with the understanding of nanoparticles design. Here, we discussed the ocular transport of topically applied drug, different barriers in its path and how the nanoparticles as drug carriers can improve the drug delivery to the eyes.

Keywords Ocular drug delivery · Topical administration · Drug transport barriers · Controlled drug release · Nanotechnology · Targeting

---

A. Zafar
Department of Pharmaceutics, Faculty of Pharmacy, Jamia Hamdard University, New Delhi 110062, India

J. Ahmad
Department of Pharmaceutics, National Institute of Pharmaceutical Education and Research (NIPER), Raebareli 229010, UP, India

R.T. Addo
Department of Pharmaceutical Sciences, School of Pharmacy, Union University, Jackson, TN, USA

S. Akhter (✉)
LE STUDIUM® Loire Valley Institute for Advanced Studies, Centre-Val de Loire region, Orléans, France
e-mail: sohailakhtermph@gmail.com

S. Akhter
Nucleic Acids Transfer by Non-Viral Methods, Centre de Biophysique Moléculaire, CNRS, UPR4301 Orléans, France

R.T. Addo (ed.), *Ocular Drug Delivery: Advances, Challenges and Applications*,
DOI 10.1007/978-3-319-47691-9_9

Introduction

The development of drug delivery approach for the transportation of drug in a bioavailable and safe manner to the target site is now becoming an exceedingly important area of biopharmaceutical research. To be sure, a large number of novel drug delivery technologies surface every year and every segment of the body part has been attempted as a potential target for the site of action.

Ophthalmic drug delivery is one of the most interesting and challenging endeavors facing the pharmaceutical scientist. The anatomy, physiology, and biochemistry of the eye render this organ highly impervious to foreign substances. This appears to be quite challenging for drug delivery scientist to circumvent the protective barriers of the eye without causing permanent tissue damage. Development of newer, more sensitive diagnostic techniques and novel therapeutic agents continued to provide ocular delivery systems with high therapeutic efficacy.

The goal of pharmacotherapeutics is to treat a disease in a consistent and predictable fashion. An assumption is made that a correlation exists between the concentration of a drug at its intended site of action and the resulting pharmacological effect. The specific aim of designing a therapeutic system is to achieve an optimal concentration of a drug at the active site for the appropriate duration. Ocular disposition and elimination of a therapeutic agent are dependent upon its physicochemical properties as well as the relevant ocular anatomy and physiology (Hanrahan et al. 2012) for treating anterior ocular disorders, topical administration into the conjunctival cul-de-sac is usually the preferred route of delivery. A successful design of a drug delivery system, therefore, requires an integrated knowledge of the drug molecule and the constraints offered by the ocular route of administration.

Conventional Approach for Ocular Drug Delivery

Medication is applied to the surface of the eye for two purposes: to treat the outside of the eye for such infections as conjunctivitis, blepharitis, keratitis sicca, or to provide intraocular treatment through the cornea for diseases such as glaucoma or uveitis. Most ocular diseases are treated with topical application of solutions administered as eye drops. These conventional dosage forms account for nearly 90 % of the currently accessible marketed formulations. Eye drops used for soluble drug, require frequent instillations of highly concentrated solutions. The practical reasons for selecting solutions are the generally favorable cost advantage, the greater simplicity of formulation development and production and the good acceptance by patients despite a little blurring (Fitzgerald and Wilson 1994; Araujo et al. 2009; Singh et al. 2011). In that, most of the drugs are washed from the eye by various mechanisms like lacrimation, tear dilution, and tear turnover. Moreover, the relatively impermeable corneal barrier restricts the entry of foreign substances. As a result, less than 5 % of administered drug penetrates the cornea and reaches intraocular tissues. Many researchers have made significant efforts to improve the ocular bioavailability by increasing the corneal residence/contact time by using novel delivery system such as aqueous eye drop with increased viscosity (Sechoy et al. 2000; Tissie et al. 2002), colloidal particles (De Campos et al. 2004; Sahoo et al. 2008; Gupta et al. 2010; Wadhwa et al. 2010), in situ gel system, dendrimers (Vandamme and Brobeck 2005), Ocular inserts (Ding 1998), collagen shields (Hill et al. 1993) and ion pair (Higashiyama et al. 2006).

Drug Transport Mechanism and Barriers in Ocular Drug Delivery

The eyes are among the most readily accessible organs in terms of location in the body and topical delivery into the cul-de-sac is, by far, the most common route of ocular drug delivery. Absorption from this site may be corneal or non-corneal. The so-called non-corneal route of absorption involves penetration across the sclera and conjunctiva into the intraocular tissues. This mechanism of absorption is usually non-productive, as drug penetrating the surface of the eye beyond the corneal-scleral limbus is taken up by the local capillary beds and removed to the general circulation. This non-corneal absorption, in general, precludes the entry into the aqueous humor. But drug delivery to eye tissues is particularly problematic due to the presence of ocular barriers like tear film, cornea, conjunctiva and sclera (Fig. 9.1a, d).

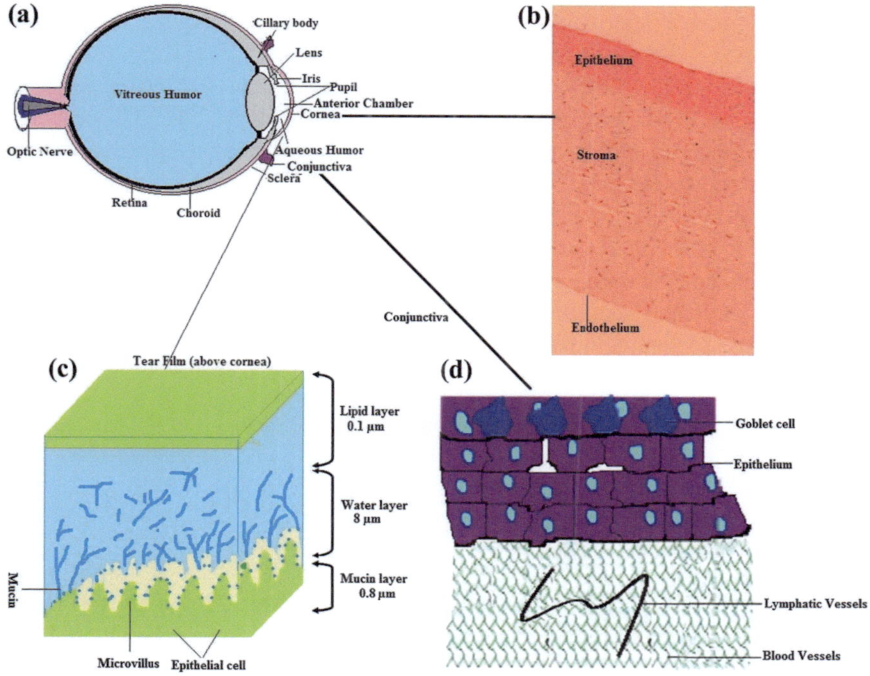

Fig. 9.1 Anatomical structure of human eye and barriers in ocular drug delivery. **a** Human eye. **b** Cornea. **c** Tear film. **d** Conjunctiva

Tear Film

The tear (precorneal film) film is the liquid layer bathing the cornea and conjunctiva. The film thickness is reported to be about 3–10 μm (Robinson 1993; King-Smith et al. 2000; Szczesna et al. 2006, 2007; Szczesna-Iskander and Iskander 2012; Dhubhghaill et al. 2012). The tear film consists of a three-layered structure comprising a lipid (oil) layer, an aqueous (water) layer and a mucous layer over the corneal epithelium as shown in Fig. 9.1c. The lipid layer is the outermost surface and polishes the corneal surface. It is produced by the meibomian glands and provides a smooth tear surface and retarding the rate of tear evaporation from the cornea. Mechanically traps and flushes out foreign bodies and chemicals contain bacteriostatic substances that inhibit the growth of microorganisms. The aqueous layer produced by the lachrymal gland is composed of water, proteins, and other substances, such as lipocalin, lactoferrin, lysozyme, and lacritin. This layer is responsible for control of infection, osmotic balance and water promoting spreading of the tear film. The inner-most layer of the tear film is mucus layer. It is produced by goblet cells of the conjunctiva and acts as a hydrophilic layer (water soluble) and serves as an anchor for the tear film and helps it adhere to the eye. The tear film creates a smooth surface for light to pass through the eye, nourishes the front of the eye, and provides protection from infection (Wolff 1946, 1954; King-Smith et al. 2000; Rathore and Nema 2009; Montes et al. 2010; Stahl et al. 2012). Tear film which reduces the effective concentration of the administrated drugs due to dilution by the tear turnover (approximately 1 μL/min), accelerated clearance, and binding of the drug molecule to the tear proteins (Schoenwald et al. 1998; King-Smith et al. 2000; Her et al. 2013). The Osmolarity of the tear film equals 310–350 mOsm/kg in normal eyes and is adjusted by the principal inorganic ions Na^+, K^+, Cl^-, $HCO3^-$, and proteins. The mean pH value of normal tears is about 7.4. Depending on age and diseases, values between 5.2 and 9.3 have been measured. Diurnal patterns of pH changes exist, with a general shift from acid to alkaline during the day. The buffer capacity of the tears is determined by bicarbonate ions, proteins, and mucin. Tears exhibit a non-Newtonian rheological behavior. The viscosity is about 3 mPas (Greaves et al. 1993; Pandit et al. 1999). The surface tension depends on the presence of soluble mucin, lipocalins, and lipids. The mean surface tension value is about 44 mN/m (Nagyova and Tiffany 1999; Tiffany 2003).

Cornea

The cornea is the transparent, dome-shaped window covering the front of the eye and one of the most densely innervated tissues on the surface of the body (Marfurt et al. 1989; Muller et al. 2003). It is a powerful refracting surface, providing 2/3 of the eye's focusing power. Because there are no blood vessels in the cornea, it is normally clear and has a shiny surface. It's approximately 11.7 mm in diameter,

500 μm thick in the center with around 700 μm at the periphery and radius of curvature of anterior surface about 7.7 mm while the radius of curvature of the globe is approximately 12 mm (Fig. 9.1b) (Maurice and Mishima 1984; Hitzenberger et al. 1994). The cornea consists of three layers; epithelium, stroma, and endothelium and passes the mechanical barrier of foreign substances. Each layer possesses a different polarity and a rate-limiting structure for drug permeation. The corneal epithelium is of lipophilic nature, and tight junctions among cells are formed to restrict paracellular drug permeation from the tear film and permeated lipophilic drugs (transcellular pathway). Hydrophilic charged cationic compounds permeate more easily through the cornea than anionic forms, because the corneal epithelium is negatively charged above its isoelectric point (Rojanasakul et al. 1992; Bourlais et al. 1998; Gipson and Argueso 2003; Baspinar et al. 2008; Araujo et al. 2009). The highly hydrated structure of the stroma acts as a barrier to permeation of lipophilic drug molecules. Corneal endothelium is the innermost monolayer of hexagonal-shaped cells and acts as a separating barrier between the stroma and aqueous humor. The endothelial junctions are leaky and facilitate the passage of macromolecules between the aqueous humor and stroma (Bourlais et al. 1998; Fischbarg 2006; Araujo et al. 2009).

Conjunctiva

The conjunctiva is the thin, vascularized mucous membrane, transparent tissue that covers the outer surface of the eye. Histologically the conjunctiva consists of non-keratinized, stratified squamous epithelium, with interspersed goblet cells. It is involved in the formation and maintenance of the tear film, In addition, conjunctiva or episclera has a rich supply of capillaries and lymphatics as shown in Fig. 9.1d (Sugar et al. 1957; Raviola 1983; Robinson 1993; Gausas et al. 1999; Singh 2003; Hosoya et al. 2005) therefore, drugs administered in the conjunctival or episcleral space may be cleared through blood and lymph. The conjunctival blood vessels do not form a tight junction barrier (King-Smith et al. 2000), which means drug molecules can enter into the blood circulation by pinocytosis and/or convective transport through paracellular pores in the vascular endothelial layer. The conjunctival lymphatics act as an efflux system for the efficient elimination of the conjunctival space. Recently, it has been reported that at least 10 % of a small molecular weight hydrophilic model compound (sodium fluorescein), administered in the subconjunctival space, was eliminated via the lymphatics within the first hour in rat eyes (Lee et al. 2010). Therefore, drugs transported by lymphatics in conjunction with the elimination by blood circulation can contribute to systemic exposure, since the interstitial fluid is returned to the systemic circulation after filtration through lymph nodes.

Sclera

The sclera (white of the eye) is white, fibrous collagen tissues, and has three layers (anterior to posterior): episclera, scleral stroma, and lamina fusca and continued to cornea anteriorly. Sclera provides the structural integrity that defines the shape and length of the eye (Newell 1993; Ludwig 2005; Pescina et al. 2011). Scleral permeability has been shown to have a strong dependence on the molecular radius. Scleral permeability decreases roughly exponentially with molecular radius. Additionally, the posterior sclera is composed of a looser weave of collagen fibers than the anterior sclera (Miao et al. 2013), and the human sclera is relatively thick near the limbus (0.53 $\pm$ 0.14 mm), thin at the equator (0.39 $\pm$ 0.17 mm) and much thicker near the optic nerve (0.9–1.0 mm). Thus, the ideal location for transscleral drug delivery is near the equator at 12–17 mm posterior to the corneoscleral limbus (Myles et al. 2005; Qi et al. 2013). Hydrophobicity of drugs affects scleral permeability; an increase of lipophilicity shows lower permeability; and hydrophilic drugs may diffuse through the aqueous medium of proteoglycans in the fiber matrix pores more easily than lipophilic drugs (Maurice and Mishima 1984; Cruysberg et al. 2002; Wen et al. 2013).

Pharmacokinetic Consideration in Ocular Drug Delivery

The most significant reason for not conducting ocular pharmacokinetic studies in the human eye is the inability to sample tissues or fluids from the intact eye without risking pain and/or injury. Although the rabbit eye is useful in predicting human ocular toxicities (McDonald and Shadduck 1977), the eyes of each species are dissimilar in anatomy and physiology (Table 9.1) such that predicting human ocular pharmacokinetics from rabbit data may not be very precise for certain drugs.

Table 9.1 Anatomical and physiological difference in human eye and New Zealand rabbit

Pharmacokinetic factor	Human eye	Rabbit eye
Tear volume (μL)	7.5	7.0–30.0
Tear turnover rate (μL/min)	0.5–2.2[a]	0.6–0.8
Spontaneous blinking rate (times/min)	15	4–5
Nictitating membrane	Absent	Present
pH of tears	7.14–7.82	7.14–7.82
Milliosmolarity of tears (mOsm/L)	305	305
Corneal thickness (mm)	0.52	0.40
Corneal diameter (mm)	15	12
Aqueous humor volume (μL)	310	310
Aqueous humor turnover rate (μL)	1.53	1.53

[a]Range depending on blinking rate and conjunctival sac volume

After topical administration of an ophthalmic drug solution, the drug is firstly mixed with the lachrymal fluid. The contact time of drug with ocular tissues is relatively short (1–2 min) because of the permanent production of lachrymal fluid (0.5–2.2/μl/min). Then, approximately half of the drug flows through the upper canaliculus and the other half, through the lower canaliculus into the lachrymal sac, which opens into the nasolacrimal duct. Drainage of lachrymal fluid during blinking (every 12 s) towards the nasolacrimal duct induces a rapid elimination of conventional dosage forms (Ahmed and Patton 1985, 1987). The drug is absorbed into the retina-choroid via a corneal or scleroconjunctival route; the iris and ciliary body are presumably supplied via both the transcorneal and the extracorneal pathways.

Drugs penetrate across the corneal epithelium via the transcellular or paracellular pathway. Lipophilic drugs prefer the transcellular route, while hydrophilic drugs penetrate primarily through the paracellular pathway, which involves passive or altered diffusion through intercellular spaces. The transcorneal penetration appears to be hindered by the binding of the drug to the corneal tissues. The cornea may act as a drug reservoir, slowly releasing the drug into the aqueous humor, where levels decrease very slowly.

Then, drugs are distributed from the aqueous humor to the intraocular tissues, i.e., iris-ciliary body, lens, vitreous and choroid-retina and eliminated mainly via aqueous humor turnover and venous blood flow in the anterior uvea (Fig. 9.2). It is suggested that ocular penetration via the scleroconjunctival route is more rapid (for a hydrophilic drug) than via the transcorneal route (Worakul and Robinson 1997). Both transconjunctival absorption and transnasal absorption after drainage via the nasolacrimal duct are generally undesirable, not only because of the loss of active ingredient into the systemic circulation but also because of possible side-effects, for instance the effects on the heart when beta-blockers are administered for the treatment of wide-angle glaucoma (Schoenwald 1990; Jtirviner et al. 1995; Meseguer et al. 1996).

Formulation Approach for Ocular Drug Delivery

Mainly two types of approaches i.e., conventional and controlled drug delivery approaches. Conventional drug delivery approaches such as solutions, suspensions, and ointments, account for almost 90 % of the currently accessible ophthalmic formulations on the market (Lang 1995; Bourlais et al. 1998). They offer some advantages such as their ease of administration by the patient, ease of preparation and low production costs. However, there are also significant disadvantages, especially with the use of conventional solutions, including the very short contact time with the ocular surface and the fast nasolacrimal drainage, both leading to poor bioavailability of the drug. Nevertheless, conventional eye drops remain the most commonly used dosage forms in ocular delivery. The various drawbacks of conventional ocular drug delivery are described in Table 9.2.

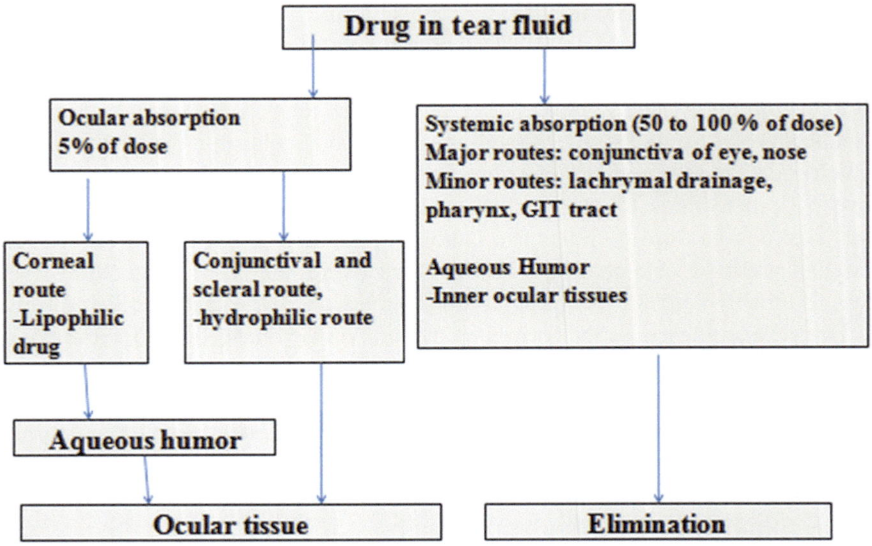

Fig. 9.2 Systemic representation of drug distribution after topical instillation in human eye

Table 9.2 Disadvantages of the conventional ocular delivery system over controlled or novel drug delivery approaches such as colloidal, viscosity enhancing formulation, in situ gel system and ocular insert

Solution	Suspensions	Ointment
• Rapid pre-corneal elimination • Loss of drug by drainage • No sustained action	• Drug properties decide performance • Loss of both solution and suspended solid	• Sticking of eyelids • Poor patient compliance • Blurred vision • No true sustained effect • Drug choice limited by partition co-efficient

Advantages of Modified Ocular Drug Delivery (Wilson 2004; Vyas and Khar 2008)

- To circumvent the protective barriers like drainage, lacrimation and diversion of exogenous chemicals into the systemic circulation by the conjunctiva
- To increase the ocular bioavailability of the drug by increasing corneal contact time. This can be achieved by effective coating or adherence to corneal surface so that the released drug effectively reaches the anterior chamber.
- To overcome the side effects of pulsed dosing produced by conventional systems.
- To provide comfort and compliance to the patient and yet improve the therapeutic performance of the drug over conventional systems.

- To provide sustained and controlled drug delivery.
- To provide targeting within the ocular globe, so as to prevent the loss to other organs.

Colloidal Carrier Exploited for Ocular Drug Delivery

Colloidal delivery systems, particularly nanoparticulate system are widely employed for the treatment of ocular disease. It would prolong the ocular retention of the drug, allowing the drug to remain in contact with the cornea for a longer duration, sustained release and thus increasing bioavailability. These delivery systems include liposomes, microemulsions/nanoemulsions, and nanoparticulate system etc.

Microemulsions/Nanoemulsions

Microemulsions/nanoemulsion (ME/NE) are of interest to the pharmaceutical scientists as promising drug delivery vehicles due to their, small size, simple and inexpensive preparation, and their sterilization easily by filtration (Soukharev and Wojciechowska 2005), low viscosity, greater ability as drug carrier, absorption promoter, and low surface tension of microemulsions, show spreading on the cornea mixing with the precorneal film constituents (Fialho 2004), long term stability, low toxicity and irritancy, considerable capacity for solubilisation of a variety of drug molecules (lipophilic and hydrophilic) and great potential in bioavailability improvement (Fialho 2004; Soukharev and Wojciechowska 2005; Gupta and Moulik 2008; Djekic et al. 2008; Kesavan et al. 2013). It consists of water, oil and a mixture of surfactants and co-surfactant, making a homogeneous, optically isotropic and thermodynamically stable and a small droplet size in the dispersed phase (<1.0 mm) (Soukharev and Wojciechowska 2005; Djekic et al. 2008).

The penetration enhancing the property of surfactant and co-surfactant of microemulsions increases the drug permeability (drug uptake) and facilitates the passage of drug by the corneal membrane. Moreover, microemulsion/nanoemulsion achieve sustained release of a drug applied to the cornea and higher penetration into the deeper layers of the ocular structure and the aqueous humor than the eye drops. Due to prolonged release of drug from microemulsions, they find an important role in ocular delivery which can decrease the frequency of application.

Kesavan and coworkers developed microemulsions of dexamethasone and found that small stable globule size, acceptable physicochemical behavior, good mucoadhesive properties, and ability to enhance bioavailability through its longer precorneal residence time sustained the release of the drug (Kesavan et al. 2013).

Ammar and co-workers described dilutable nanoemulsions of dorzolamide. They developed nanoemulsion of dorzolamide hydrochloride using different oils, surfactants, and co-surfactants by applying pseudo ternary-phase diagrams. These nanoemulsions showed acceptable physicochemical properties and exhibited slow drug release and no sign of inflammation on Draize rabbit. Biological evaluation of dorzolamide hydrochloride nanoemulsions on normotensive albino rabbits indicated that these products had higher therapeutic efficacy, faster onset of action, and prolonged effect relative to either drug solution or the market product (Ammar et al. 2009).

The high-level indomethacin in inner ocular structure, aqueous humor in rabbit eyes was achieved upon topical instillation of chitosan nanoemulsion as compared with indomethacin drug solution. The chitosan nanoemulsion prepared were able to make contact intimately with the cornea thus giving a slow continuing drug release with long-term drug intensity, thus enhancing delivery to both internal and external ophthalmic tissues (Badawi et al. 2008; Addo et al. 2010, 2015). The microemulsions system exploited for ocular delivery of various drugs is shown in Table 9.3

Tayel and co-workers developed controlled-release in situ ocular drug-loaded nanoemulsion using pseudo ternary phase diagrams technique (NE) gels. The formulation contained isopropyl myristate/Miglyol 812), surfactants (Tween 80/Cremophor EL), co-surfactant (polyethylene glycol 400) and water and gellan gum solution (0.2 %, w/w). The optimized formulation was evaluated in vitro and in vivo and showed transparency, rheological behavior, mucoadhesive force, sustained drug release. On instillation in rabbit eye, it showed good consistency, thermodynamically stable and no irritation and histopathological assessment of ocular irritation. The gel had significantly ($P < 0.01$) higher C_{max}, delayed T_{max},

Table 9.3 Microemulsion system exploited for ocular drug delivery

Drug	Formulation	Outcomes	Reference
Dexamethasone	Chitosan-coated Microemulsion	Significant Enhanced bioavailability	Kesavan et al. (2013)
Everolimus	Microemulsion	Microemulsion was Stable, 8.64 ng/mL drug reached in 30 min	Baspinar et al. (2008)
Vitamin A palmitate	Microemulsion	Showed better wettability and longer ocular retention	Ma et al. (2008)
Chloramphenicol	Microemulsion	Bioavailability increased compared to eye drops	Lv et al. (2005)
Dexamethasone	Microemulsion	It increased the AUC two-fold higher than conventional eye drops	Fialho (2004)
Pilocarpine	Microemulsion	Significantly Decreased intraocular pressure by 25 %	Garty and Lusky (1994)
Timolol	Microemulsion	Significantly increased AUC 3.5 fold than timolol eye drops in aqueous humor	Gasco et al. (1989)

prolonged mean residence time and increased bioavailability as compared to terbinafine hydrochloride eye drops and maintained effective aqueous humor concentrations (Tayel et al. 2013).

Nanoparticles (Nps)

Nanoparticles (NPs) are one of the most studied colloidal systems with the object of improving targeting of the drug to organs and increasing drug bioavailability across biological membranes. NPs are sub-microscopic, a colloidal system consisting of macromolecular substances that vary in size from 1 to 1000 nm. In NPs the drug may be dissolved, entrapped, adsorbed, attached or encapsulated into the polymer matrix.

Depending on the method of preparation, it can be classified into two groups: NPs (nanospheres) and nanocapsules and have different release profile of drug (Sahoo and Labhasetwar 2003; Omidi et al. 2012).

Nanospheres are small solid spheres constituting of a dense solid polymeric network having a large surface area (Omidi et al. 2012). Drugs can either be incorporated into the matrix system or adsorbed on the surface of the nanospheres. On the other hand, nanocapsules have a small central cavity (oily droplet) surrounded by a polymeric membrane and drug can either be incorporated into the matrix system or adsorbed on the surface (Fig. 9.3).

NPs represent promising drug carriers for ophthalmic applications. After optimal binding of these particles, the drug absorption in the eye is enhanced significantly in comparison to eye drop solutions owing to the much slower ocular elimination rate of particles. Smaller particles are better tolerated by the patients than larger particles; therefore NPs may comfortably be used for prolonged action ophthalmic delivery systems.

Different polymers can be used to fabricate NPs such as biodegradable polymers like polylactide (PLAs), poly (D, L-lactides), polylactic-co-glycolic acid, E-caprolactone, (Fessi et al. 1989; Kumari et al. 2010; Aksungur et al. 2011; Gupta et al. 2011), polyacrylamide, poly cyanoacrylate and polymethylmethacrylate (Zimmer et al. 1991; Wenger et al. 2011) and natural polymers like CS (Jain et al. 2014), gelatin (Vandervoort and Ludwig 2004), sodium alginate (Zhu et al. 2012), albumin (Zimmer et al. 1994; Merodio et al. 2002) and tamarind kernel polysaccharide (Kaur et al. 2012) can be used effectively for efficient drug delivery to the ocular tissues. Many researchers conducted the study on nanoparticulate drug delivery system and found that it prevents the degradation of the drug in an ocular environment and release of drug over an extended period of time which gives the desired effect (Gupta et al. 2010; Javadzadeh et al. 2010) (Table 9.4).

Başaran and coworkers showed in an in vivo study that CS-NPs of cyclosporine prolonged release of active agent due to the positive charge of CS. This may be attributed to enhanced residence time at the corneal and conjunctival surfaces (Başaran et al. 2014).

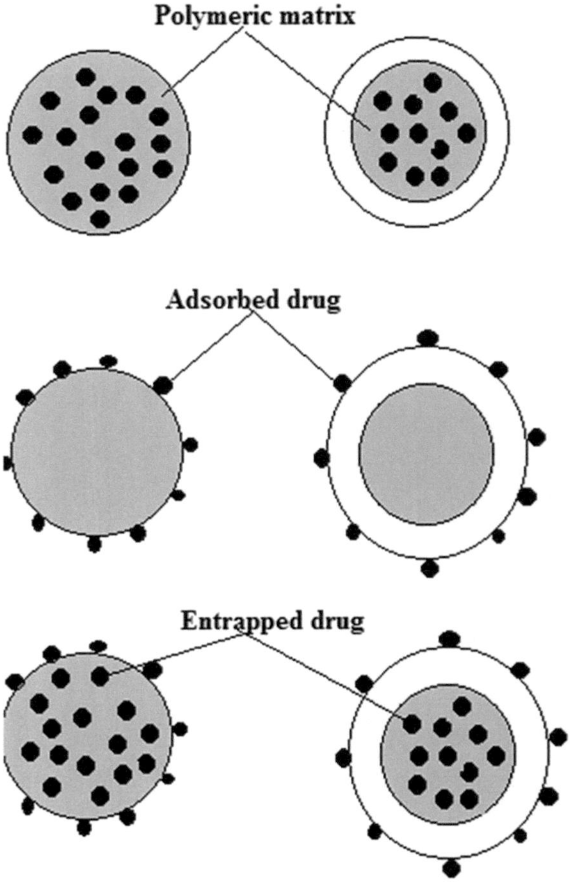

Fig. 9.3 Different type of NPs exploited in ocular drug delivery

Sabzevari and coworkers studied the biodegradable poly β-amino ester-NPs of triamcinolone acetonide of anti-inflammatory effects in the rabbit eye. It was concluded that polymeric NPs of triamcinolone acetonide will provide good anti-inflammatory effects as subconjunctival injection method and are better compared to other drug delivery systems (Sabzevari et al. 2013).

Wadhwa and coworkers reported that hyaluronic acid (HA) gave the synergistic effect for mucoadhesion in association with dorzolamide hydrochloride or timolol maleate loaded CS-NPs and revealed that CS-HA-NPs show a higher reduction in intraocular pressure level as compared to pure drug solution (Wadhwa et al. 2010).

Agnihotri and Vavia evaluated diclofenac sodium–loaded PLGA-NPs for ocular use and found good biocompatibility with the eye (Agnihotri and Vavia 2009).

Jwala and coworkers studied the PLGA-NPs delivery of acyclovir prodrug for the treatment of ocular herpes keratitis. It was found that NPs exhibited a biphasic release behavior i.e., initial burst phase followed by sustained release because PLGA-NPs may retard the degradation of the prodrug in the precorneal area and

Table 9.4 Colloidal system exploited for ocular delivery of different drugs

Polymer	Drug	Type of formulation	Outcomes	References
Chitosan	Betaxolol	Nanoparticles	Significant reduction in intraocular pressure	Jain et al. (2013)
Chitosan	Carteolol	Nanoparticles	Significant reduction in intraocular pressure because of increase in corneal retention time	Ameeduzzafar et al. (2014)
Chitosan, carbopol	Dorzolamide	Nano gel	In situ gel of dorzolamide hydrochloride, Nanogel showed good ocular retention over corneal surface	Katiyar et al. (2014)
Dorzolamide-loaded PLGA/vitamin E TPGS	Dorzolamide	Nanoparticles	Pharmacoscintigraphic studies revealed the reduced corneal clearance, as well as nasolacrimal drainage in comparison to drug solution. Significant reduction in intraocular pressure	Warsi et al. (2014)
Cyclodextrin nanoparticle suspension	Dorzolamide	Nanoparticle suspension	Dorzolamide γ-cyclodextrin nanoparticle eye drops are no more locally toxic or irritating to the eye than Trusop.	Jóhannesson et al. (2014)
Hyaluronic liposome	Doxorubicin (DOX)	Mucoadhesive liposomes	HA-modified liposomes had the longest retention time, following with naked liposomes. Significantly, the area under the curve of the aqueous humor concentration-time profiles of DOX liposomes was found to be 1.7-fold higher than that of DOX solution	Lin et al. (2016)
Hyaluronic liposome	Tacrolimus	Hyaluronic acid-coated niosomes	The hybrid system showed higher ocular delivery on mucoadhesion, precorneal retention, and aqueous humor pharmacokinetics and transcorneal permeability	Zeng et al. (2016)
Hyaluronic acid-chitosan	Dexamethasone	Nanoparticles	Increased tear concentration and increased bioavailability due to their prolonged precorneal-retention because of highly mucoadhesive characteristics of CS and HA.	Kalam (2016)

(continued)

Table 9.4 (continued)

Polymer	Drug	Type of formulation	Outcomes	References
Gelucire 44/14 and amphipathic octadecyl-quaternized carboxymethyl chitosan	Tetrandrine	Liquid crystalline nanoparticles	LCNP system could be a promising method for increasing the ocular bioavailability of TET by enhancing its retention time and permeation into the cornea	Liu et al. (2016)
Glyceryl monoolein (GMO) and water in the presence of stabilizer Poloxamer 407	Pilocarpine	Liquid crystalline nanoparticles	Showed prolonged effect on decreasing intraocular pressure (IOP) than commercial drug and physiological saline	Li et al. (2013)
Glyceryl monooleate/poloxamer 407	Cyclosporine	Liquid crystalline nanoparticles	Liquid crystalline nanoparticles exhibited excellent ocular tolerance in the ocular irritation test	Chen et al. (2012)
Chitosan, PLGA	Fluocinolone Acetonide	Nanoparticles	Pharmacokinetic studies on Albino rabbit's eyes using HPLC indicated that the prepared novel chitosan-coated PLGA nanoparticles subjected to separation by filtration showed rapid and extended drug delivery to the eye	Salama et al. (2015)
Poly(lactic-co-glycolic) acid	Carprofen	Nanoparticles	In vivo ocular anti-inflammatory efficacy, the test confirmed an optimal efficacy of NPs and its potential application in eye surgery	Parra et al. (2015)
Hyaluronic acid + CS	Dorzolamide hydrochloride (DH) or timolol maleate	NPs	Significant reduction in intraocular pressure due to enhanced ocular residence time	Wadhwa et al. (2010)
Lipid	Diclofenac sodium	liposomes	Diclofenac sodium cationic liposomes increased the corneal contact time, enhance the corneal permeability of diclofenac sodium and improve its ocular bioavailability	Sun et al. (2006)

(continued)

Table 9.4 (continued)

Polymer	Drug	Type of formulation	Outcomes	References
Sodium alginate-CS	5-fluorouracil (5-FU)	NPs	Bioavailability NPs was significantly higher than 5-FU solution in aqueous humor of rabbit eye	Nagarwal et al. (2012)
Albumin	Ganciclovir (GCV)	NPs	Showed prolonged retention over corneal surface having sustained release over a day	Merodio et al. (2002)
Albumin	Ganciclovir (GCV)	NPs	Increased antiviral activity against human cytomegalovirus (HCMV) infection	Irache et al. (2005)
Non-ionic surfactant	Timolol maleate	Niosome	Formulation showed 1.7 times higher aqueous humor concentration of drug as compared to the timolol maleate solution	Kaur et al. (2010)
Span 60	Acetazolamide	Niosome	Permeability increase and increased in aqueous humor drug concentration	Aggarwal et al. (2007)

further sustain the release in deep corneal tissues. Dispersion of NPs in thermosensitive gels completely eliminated the burst release phase (Jwala et al. 2011).

Basaran and co-workers developed cyclosporine NPs for ocular delivery by spray drying technique by using different grade (molecular weight) of CS and characterized in vitro and in vivo. In vitro study showed sustained release. The in vivo study was done using sheep and sample was analyzed by enzyme immune assay. It showed that prolonged release of active agent from positively charged CS formulations after 72 h of study in both aqueous and vitreous humor samples. This was attributed to increasing in the corneal and conjunctival surfaces residence, so it was concluded that CS possess bioadhesive and permeation enhancing property (Başaran et al. 2014).

Tayel and coworkers formulated terbinafine hydrochloride NPs by positively charged Eudragit® RS 100 and CS polymer in a different ratio by nanoprecipitation technique using Soybean lecithin (1 %, w/v) and Pluronic® F68 as surfactant and stabilizer. The NPs have small particles size, zeta potential (73.29–320.15 nm and +20.51 to +40.32 mV respectively) and sustained release profile in simulated tear fluid (pH 7.4). The NPs were physically stable at 4 and 25 °C. NP suspension showed significantly ($P < 0.05$) increased drug mean residence time and improved its ocular bioavailability; 1.657-fold in rabbits aqueous humor as compared to terbinafine hydrochloride eye drops (0.25 %, w/v). To enhance the permeation and residence time of the formulation, by overcoming the limitations associated with protective ocular barriers (Tayel et al. 2013).

Pathak and co-workers evaluated novel pH-triggered nano-emulsified in situ gel (NE-ISG) for ophthalmic delivery of fluconazole. Pseudoternary phase diagrams were constructed using capmul MCM (oil phase), tween 80 (surfactant) and transcutol P (cosurfactant) to identify the nanoemulsion region and formation of spontaneous emulsification. The formulations were characterized by permeation, corneal irritation, and corneal toxicity. The optimized nanoemulsion in situ gels showed 3-fold (337.67 $\mu g/cm^2$) higher Ex vivo transcorneal permeation than commercial eye drops (112.92 $\mu g/cm^2$). Formulation did not show any visual signs of tissue damage. Hence it was concluded that nanoemulsion gel may offer a more intensive treatment of ocular fungal infections due to higher permeation, prolonged precorneal residence time and sustained drug release along with higher in vitro efficacy, safety and greater patient compliance (Pathak et al. 2013).

Mohammed and coworkers developed and studied the fluconazole loaded chitin nanogels (Flu-CNGs) for the treatment of corneal fungal infections. Flu-CNGs have controlled release pattern at a prolonged period of time. Flu-CNGs are hemocompatible, cytocompatible and also showed very good cell uptake in human dermal fibroblast cells and penetration to the deeper sections of the porcine cornea with no signs of destruction or inflammation to corneal cells (Mohammed et al. 2013).

Du Toit and coworkers developed and compared two specific embodiments of an ocular nanosystem (NS) i.e., CS-poly (ε-caprolactone), and the composite lipoidal-polymeric NS for intelligent treatment of inflammatory disorders (specifically ocular afflictions). The anti-inflammatory efficacy, demonstrated by a decrease in 4-chloro-7-nitrobenzo-2-oxa-1, 3-diazole complex formation (0.0031 Vs 0.0023 mmol/L) and decrease in NFκB formation (decrease in relative optical density of 0.2027 vs. 0.2420). It was found that indomethacin lipoidal-polymeric NS has enhanced efficacy in terms of tissue permeation, cell uptake, and anti-inflammatory activity (Du Toit et al. 2013).

Zhou and co-worker developed a self-aggregated nanoparticulate vehicle using an amphiphilic poly (lactic acid)-grafted-CS (PLA-g-CS) copolymer and to evaluate its potential for ocular delivery of amphotericin B. The self-aggregated PLA-g-CS NPs had a core-shell structure with an average particle size of approximately 200 nm and zeta potentials higher than +30 mV with high encapsulation efficiency. The formulation showed sustained release profile and no sign of irritation after instillation in rabbit eye was scored. The significant antifungal activity was observed when compared with amphotericin B *against Candida albicans*. The in vivo ocular pharmacokinetic study suggested that the PLA-g-CS-NPs have the advantage of prolonging residence time at the ocular surface. The corneal penetration study showed that the PLA-g-CS NPs could penetrate into the cornea (Zhou et al. 2013).

Kaur and co-workers developed the spanlastics which consisted of spans and an edge activator prepared by the ether injection method and characterized by size, shape, and the number of vesicles/ml by optical microscopy, Entrapment efficiency, ex vivo corneal permeability study and MTT assay. Spanlastics formulation is smaller in size (3 times) and showed a better permeation in comparison to a corresponding niosomal formulation and 3-fold increase compared to the marketed formulation due to its elastic nature. The developed system is novel and provides an effective and safe formulation of fluconazole (Kaur et al. 2012).

Başaran and co-workers developed cyclosporine A NPs using cationic Eudragit RS for ocular application. The formulation was characterized in vitro and in vivo. The NPs showed the extended release of incorporated drug. In vivo study showed prolonged residence time off, cyclosporine A in the deeper layers (vitreous humor) of the eye due the positively charged nature of Eudragit RS (Başaran et al. 2011).

Singh and coworkers developed CS-NPs of brimonidine tartrate (BT) for ocular delivery and characterized by TEM, SEM, particle size, and polydispersity index (PI), DSC, IR, and entrapment efficiency which gave an insight of physicochemical interaction that influenced the CS-NPs formation and entrapment of BT. In vitro release showed sustained release. In vivo studies confirmed a significant sustained effect BT-NPs were compared with conventional eye drops. It was concluded that BT loaded CS-NPs could help to reduce dosing frequency by sustained drug release in the treatment of glaucoma (Singh et al. 2011).

Liposomes

Liposomes were first introduced as drug delivery carriers by Bangham et al. (1965). They bilayer lipid vesicles composed of altering aqueous compartments enclosed by lipid bilayers (mainly phospholipids and cholesterol) of natural and synthetic origin, and usually within the size range of 10 nm to 1 μm or greater (Meisner and Mezei 1995; Mishra et al. 2011; Honda et al. 2013). Liposomes have advantages like, improved bioavailability of ophthalmic drugs after topical administration (Monem et al. 2000; Kaur and Kanwar 2002; Wadhwa et al. 2009), no toxicity and low antigenicity (Rooijen and Nieuwmegen 1980) stable, biocompatible, biodegradable and metabolize in vivo encapsulating both lipophilic, hydrophilic and amphiphilic molecules (Klibanov et al.1990; Abrishami et al. 2009) hydrophilic drugs are entrapped in the aqueous layer, while hydrophobic drugs are stuck in the lipid bilayers. Depending on their size and number of bilayers, liposomes can be classified as small unilamellar vesicles (SUVs, 20 nm to ~200 nm), giant unilamellar vesicles (GUVs, >1 μm), and large unilamellar vesicles (LUVs, 200 nm to ~1 μm) and, Multilamellar vesicles (MLV, >0.5 μm) (Jesorka and Orwar 2008). Unilamellar vesicles are composed of single layer of lipid such as lecithin or phosphatidylglycerol encapsulating aqueous interior core whereas Multilamellar vesicle is composed of various layers of lipid bilayers (Kaur et al. 2004; Mainardes et al. 2005; Shen and Tu 2007; Elbayoumi and Torchilin 2010; Dai et al. 2013). Liposome can also be divided into three categories i.e., negatively charged, positively charged and neutral liposomes. The charge present on the liposome is due to charged ingredient used in formulation development. Many researchers investigated that positively charged liposomes exhibited prolonged precorneal retention than negative and neutral charge liposome, due to electrostatic interaction with the negatively charged corneal epithelium, thus enhance drug absorption (Monem et al. 2000; Abuzaid et al. 2003; Danion et al. 2007; Dai et al. 2013).

Hitoshi Sasaki and co-workers reported the surface modification of liposome with poly-L-lysine to enhance the efficiency of coumarin-6 as a model drug and fluorescent marker to the retina (delivery to the posterior segment of the eye). It was found that surface modification with low molecular weight poly-L-lysine significantly increase the delivery as compared to high molecular weight poly-L-lysine, because aggregation of surface-modified liposomes, increased particle size hampered distribution to inner ocular tissue (Sasaki et al. 2013).

Hathout and co-workers reported acetazolamide liposome showed that positively charged and neutral liposomes exhibited greater lowering in IOP and a more prolonged effect than the negatively charged ones. The findings suggest that liposomes enhance corneal penetration of drug by being adsorbed onto the corneal surface, with direct transfer of drug from liposomal to epithelial cell membranes (Hathout et al. 2007).

Drug loading capacity and entrapment efficiency of liposomes depend on many factors such as the size of liposomes, concentration and types of lipid used, and physicochemical properties of therapeutic agent itself. Loading capacity and entrapping efficiency are poor for SUVs in comparison to MLVs. However, LUVs provide a balance between size, loading capacity and entrapment efficiency (Jesorka and Orwar 2008; Ding et al. 2005).

Bochot and coworkers reported the phosphodiester oligonucleotide encapsulated in liposome showed sustained release in the vitreous humor, retina and choroid (37 % even after 15 days) compared with the release from the solution and in a reduced distribution to the non-targeted tissues (sclera, lens) (Bochot et al. 1998a, b).

Szulc and coworkers reported that the liposomal encapsulation of the hydrophilic drug, pilocarpine, enhanced its pharmacological effect in rabbits when using positively charged vesicles (Szulc et al. 1988).

Chetoni and coworkers compared Acyclovir containing positively charged unilamellar liposomes compared with pure acyclovir in solution and marketed acyclovir ointment containing same acyclovir concentrations (0.12 %). It was observed that the liposomal formulation have the highest drug concentration in the aqueous humor of rabbits. The in vitro release profile of liposomal formulation showed sustained release. It was concluded that positively charged liposome bind to negatively charge corneal epithelium enhances the efficacy of acyclovir. These results indicated a significant advantage of acyclovir liposome as an alternative to acyclovir ointment (Chetoni et al. 2004).

Similarly many antiviral drugs like iododeoxyuridine, ganciclovir and acyclovir-loaded in a liposome (immuno-liposome) have been reported for the treatment of ocular herpes infection (herpes simplex virus) (Norley et al. 1986).

Niosomes

Niosome is a novel nonionic surfactant, a bilayered vesicular system formed by self-assembly of hydrated surfactant. Like liposome, they deliver a drug in a controlled manner to enhance bioavailability and get therapeutic effect over a longer period of time (Mujoriya and Bodla 2011; Karim et al. 2012). But the liposome has some drawbacks like high cost and limited shelf life, and these are overcome by developing the new vesicular system niosome (Mahale et al. 2012). The vesicle suspension is a water based vehicle; this offers high patient compliance in comparison with an oily delivery system. They possess an infrastructure consisting of hydrophilic, amphiphilic and lipophilic moieties together and can entrap both hydrophilic and lipophilic drug with a wide range of solubilities. The bilayers of the niosomes protect the enclosed active pharmaceutical ingredient from the heterogeneous factors present both inside and outside the body so can be used for labile and sensitive drugs (Khan et al. 2011). Mainly nonionic surfactant preferred due to their nontoxic, stable, ability to maintain pH up to physiological pH, act as solubilizer, wetting agents, and permeability enhancer, improve the bioavailability, controlled release of drug, nonimmunogenic biodegradable and biocompatible properties. The order of toxicity potential of various surfactant are cationic > anionic > ampholytic > nonionic. It comprised of both polar and nonpolar segments and has high interfacial activity. The formation of bilayered vesicles instead of micelles is dependent on the hydrophilic–lipophilic balance (HLB), chemical structure, critical packing parameter (CPP) of the surfactant. The chain length and size of the hydrophilic head group of the nonionic surfactant affect the entrapment efficiency of the drug. Nonionic surfactants with stearyl (C_{18}) chains show higher entrapment efficiency than those with lauryl (C_{12}) chains. The Tween series of surfactants bearing a long alkyl chain and a large hydrophilic moiety in combination with cholesterol in a 1:1 ratio have the highest entrapment efficiency (Arunothayanun et al. 2000). HLB value in the range 14–17 is not suitable to produce niosomes whereas one with an HLB value of 8.6 gives niosomes with the highest entrapment efficiency. Entrapment efficiency decreases as the HLB value decreases from 8.6 to 1.7 (Lawrence et al. 1996; Biswal et al. 2008; Shahiwala et al. 2002; Kumbhar et al. 2013). For HLB > 6, cholesterol must be added to the surfactant in order to form a bilayered vesicle and for lower HLB values, cholesterol enhances the stability of vesicles. It is also seen that the addition of cholesterol enables more hydrophobic surfactants to form vesicles, suppresses the tendency of the surfactant to form aggregates, and provides greater stability to the lipid bilayers by promoting the gel-liquid transition temperature of the vesicle of water soluble drugs (Khazaeli et al. 2007). Different non-ionic surfactant used in the manufacturing of niosomes like alkyl ethers and alkyl glyceryl ethers. The component of niosomes are surfactant, ether linked surfactants i.e., polyoxyethylene 4 lauryl ether (Brij 30), polyoxyethylene acetyl ethers (Brij 58) (Carafa et al. 2002; Manosroi et al. 2003; Raymzond et al. 2006), Polyoxyethylene fatty acid esters (Brij 72) (Pardakhti et al. 2007; Balakrishnan et al. 2009), Sorbitan fatty acid esters (Guinedi et al. 2005); Cholesterol; Charge Inducers: increase the stability of the vesicles by induction of charge on the surface preventing the fusion of vesicles due to repulsive forces of the same charge and provide higher values of zeta potential, negative charge inducers are diacetyl phosphate, dihexadecyl phosphate and lipoamine acid and positive charge inducers are sterylamine and cetyl pyridinium chloride (Shan et al. 2008). Gemini and Bola surfactants also used for preparation of niosome have, non-toxic, more stable, non-irritating non-haemolytic higher solubility, lower aggregation number and have no drawback like other surfactants such as Polysorbate (Harmful to persons with Crohn's disease) (Hait and Moulik 2002; Ikeda 2003; Yan et al. 2009; Carol and Keita 2010).

Different methods are used for the preparation of niosomes: such as lipid layer hydration, reversed phase evaporation, micro-fluidization, ether injection, the bubbling of nitrogen, Handjani–Vila method, enzymatic method, single pass technique involving evaporation to produce a lipid film followed by hydration with the hydration medium.

Basha and co-workers developed span 60 based nanovesicles with edged activator Tween 80 (TW80), sodium cholate (SC) or sodium deoxycholate (SDC) of clotrimazole. They found that spherical unilamellar vesicle with 87.92 % entrapment efficiency and displayed sustained antifungal effect over 12 h against *Candida albicans*. The AUC of the optimized formulation was 3.09 times more than that of drug suspension with no sign of irritation after testing for ocular tolerance (Basha et al. 2013).

Hamdy and coworkers reported the Span 60-based niosomes for ocular delivery of naltrexone HCl (NTX) and found that 5-fold increase in NTX entrapment efficiency (EE %). Ex vivo transcorneal permeation studies conducted using excised cow corneas showed that niosomes were capable of controlling NTX permeation, enhanced its corneal permeability and practically nonirritant (Hamdy et al. 2011).

Akhter and coworker developed ganciclovir mucoadhesive niosomal dispersion. It was revealed from the results that the developed formulations were nonirritant and nontoxic in nature. It was further concluded that ganciclovir nanoformulation could be utilized as a potential delivery system for treatment of ocular infections by topical instillation (Akhter et al. 2011).

Kapadia and coworkers reported niosomes incorporated in situ gel hydrogel for ocular delivery are found effective in controlling the drainage of the formulation from the ocular site and were effective in improving the formulation performance (Kapadia et al. 2009).

Aggarwal and coworkers reported that CS-coated timolol maleate (0.25 %) niosomal formulation exhibits more effect on intraocular pressure reduction with less chance of cardiovascular side effects when compared to a marketed formulation timolol solution (TMS; 0.25 %) (Aggarwal and Kaur 2005).

Ahmed and coworkers developed the non-ionic surfactant niosome delivery of acetazolamide. The results of this study showed the effect of cholesterol content, the type of surfactant and the method of preparation on the entrapment efficiency and in vitro drug release. The higher entrapment efficiency was obtained with multilamellar niosomes prepared with span 60 and cholesterol in a 7:6 molar ratio. The in vivo evaluation of acetazolamide niosomes in comparison to acetazolamide solution and niosomes showed that acetazolamide multilamellar niosomes most effectively prolonged the decrease in the IOP. Slight irritation could be observed in corneal tissues after long treatment with these niosomes which is reversible and abolished by time (Ahmed et al. 2005).

Dendrimers

Dendrimers are particularly auspicious potentially usable in numerous applications. It is a tree like highly branched synthetic nanostructured polymers with a three-dimensional structure. These very monodisperse molecules are composed of an initiator core, interior layers of repeating units, and multitudinous terminal groups. Their branched layered architectures displayed a high number of controlled terminal groups which favored biomedical applications (Bai et al. 2006; Newkome and Shreiner 2008; Kambhampati and Kannan 2013). They are classified by the number of branches and terminal groups. Various cores and units can be used, which can change the properties and shape of the dendrimer. The drug delivery can be achieved by coupling a drug through one of two approaches. Hydrophobic drugs can be complexed within the hydrophobic dendrimer interior to make them water-soluble or drugs can be covalently coupled to the surface of the dendrimer. Dendrimers can either attach to a therapeutic agent by a permanent or separable bond to the end groups or be enclosed within the dendrimer itself. Because of its multiple terminal groups and its polymer backbone, dendrimers can have multiple functionalities. In addition, dendrimers have been shown to be capable of bypassing efflux transporters and enable the efficient transport of drugs across cellular barriers (Yang and Kao 2006; Samad et al. 2009; Abdelkader et al. 2011).

Dendrimeric structures are of particular interest in the field of drug delivery due to their peculiar structural properties including controllable internal cavities bearing specific species for the encapsulation of guest drugs and external periphery with 3D multiple functional moieties for solubilisation, conjugation of bioactive compounds and targeting molecules, and recognition purposes The main successes of dendrimers resulted in their appropriate, reproducible and optimized design parameters addressing physicochemical limitations of classical drugs (e.g., solubility, specificity, stability, biodistribution and therapeutic efficiency) (Jain and Gupta 2008) and their ability to overcome biological issues to reach the right target(s) (e.g., first-pass effect, immune clearance, cell penetration, off-target interactions, etc.). Improvement of pharmacokinetic (PK) and pharmacodynamic (PD) behaviors of both drug-dendrimer conjugates and drug–dendrimer encapsulates versus drugs demonstrates their strong potentials in medicine as nano-carriers (Gajbhiye et al. 2009; Mintzer and Grinstaff 2011). The most commonly used dendrimers in nanomedicine are polyamidoamines (PAMAM), poly(L-lysine), scaffold dendrimers (PLL), polyesters (PGLSA-OH), polypropylimines (PPI), poly(2,2-bis (hydroxymethyl) propionic acid scaffold dendrimers (bis-MPA) and aminobis (methylene phosphonic acid) scaffold dendrimer.

Contemporary View: Paradigm Shift Towards Mucoadhesive Nano-System

From last 20 years, efforts have been directed in the rational design of ocular drug delivery consisting of mucoadhesive nanocarriers. Thereafter, the plan was directed to generate nanocarriers with a hydrophilic coating with the idea of improving their stability and their interaction with the mucosa (Calvo et al. 1997; Schipper et al. 1997; De Campos et al. 2003; Ludwig 2005). It was proficient via optimization of nanocarrier's ocular drug delivery to obtain long-lasting bioadhesive/residence time by the so-called 'mucoadhesive property' based on entrapment of particles in the ocular mucous layer and interaction of bioadhesive polymer chain with mucin (Henriksen et al. 1996; Du Toit et al. 2013). Maintenance of the designed nanocarriers in the ocular delivery following topical application is thus decisive to accomplish unremitting drug release and prolonged therapeutic activity. The concept of enhanced ocular resident time and bioavailability of drug through mucoadhesive nanocarriers and positively charged mucoadhesive nano-system and its interaction with the ocular surface are illustrated in Fig. 9.4.

Therefore, a design of nanocarriers from mucoadhesive materials is crucial for improved retention in the ocular cul-de-sac (Du Toit et al. 2013). Ocular mucoadhesion, exclusively, refers to the capability of certain polymers to hold on to the mucous layer casing the conjunctival and corneal surfaces of the eye by

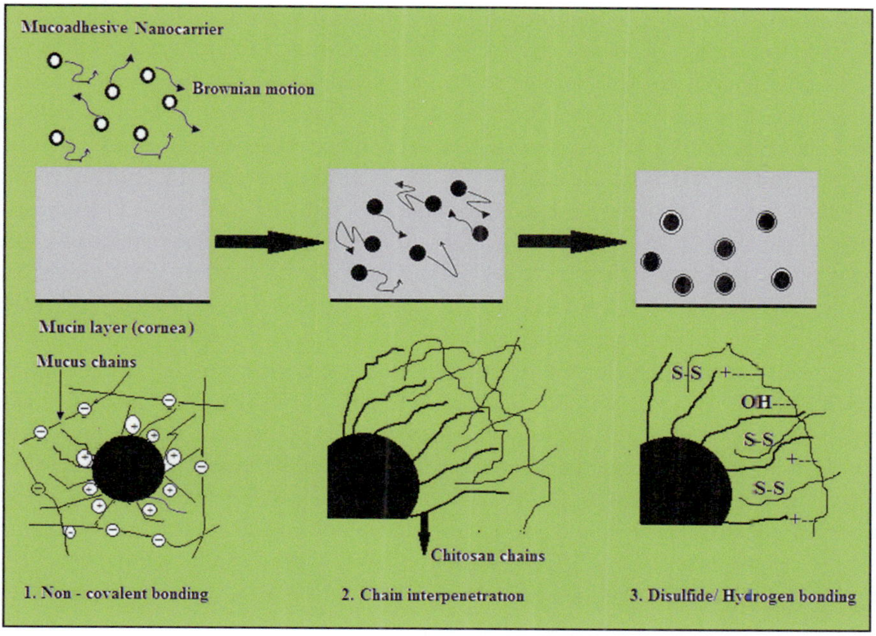

Fig. 9.4 Proposed mechanism of mucoadhesion over the corneal surface exploiting mucoadhesive and positively charged nano-system

noncovalent bonds (Round et al. 2002). Washout time for the mucoadhesive polymeric system is reduced since this depends on mucous turnover rate rather than lachrymal discharge turnover rate. A mucoadhesive polymer with plentiful hydrophilic functional group viz sulfate, hydroxyl, carboxyl, amide has found a fundamental role in ocular drug delivery system owing to their adhesion property with precorneal/conjunctival mucin layer via non-covalent bonds, and remaining in place for as long as the mucin is available there. Using this concept, various investigators planned the cationic polymer CS and CMKP as a polymer of choice because of its unique properties, including good enough biodegradability and biocompatibility permeability enhancer by affecting both paracellular and intracellular pathways of epithelial cells in a reversible manner without affecting cell viability or causing membrane wounds (Hirano et al. 1990; Takeuchi et al. 1996; Calvo et al. 1997; Felt et al. 1999; Motwani et al. 2008).

For the selection of bioadhesive polymer intended for ophthalmic drug delivery, the viscosity and wetting properties of the polymer are considered. Viscosity measures the resistance to flow which depends on upon its molecular mass, concentration, temperature and shear stress. Polymer showing non-Newtonian behaviour, when incorporated in formulation possessing pseudoplastic behaviour in which viscosity decreases with increasing shear rate (due to blinking and ocular movement), this results is significantly less resistance to blinking and demonstrates greater acceptance as compared to formulation possessing polymer exhibiting Newtonian flow, no real improvement of bioavailability in polymer showing Newtonian flow system (Schoenwald et al. 1978; Greaves et al. 1993; Ludwig and Ooteghem 1992; Van Ooteghem 1995). The mucoadhesive properties of poly-acrylic acid hydrogel and their ability to penetrate the mucin at the surface of the eye have been investigated extensively (Ponchel et al. 1987; Slovin and Robinson 1993; Duchfine et al. 1988; Meng et al. 2011). There are many mucoadhesive polymers used for designing of a colloidal carrier (Greaves et al. 1993; Govender et al. 2005) in ocular drug delivery. Various mucoadhesive polymers and their mucoadhesive strength are graded in Table 9.5.

Several other synthetic polymers have been examined for the fabrication of mucoadhesive nanocarriers for ocular delivery, for example, Pignatello and coworkers reported the formulation and evaluation of nanocarriers composed of Eudragit-RL100 with good ocular tolerance, and no inflammation or discomfort in the rabbit eye (Pignatello et al. 2002). Similarly, Barbault-Foucher and co-workers reported the hyaluronic acid (HA) colloidal system as a natural, non-irritating carrier system that showed pseudoplastic behavior with desirable ocular mucoadhesive properties (Barbault-Foucher et al. 2002).

Table 9.5 Various mucoadhesive polymers and their mucoadhesive strength

Polymer	Charge	Mucoadhesive strength
Poly (acrylic acid) (neutralized)	Anionic	+++
Carbomer (neutralized)	Anionic	+++
Hyaluronan	Anionic	+++
Chitosan	Cationic	+++
Sodium carboxymethylcellulose	Anionic	++(+)
Carboxymethyl kernel polysaccharide	Anionic	++
Poly (galacturonic acid)	Anionic	++
Sodium alginate	Anionic	++(+)
Pectin	Anionic	++(+)
Xanthan gum	Anionic	+
Xyloglucan gum	Anionic	+
Scleroglucan	Anionic	+
Poloxamer	Non-ionic	+(+)
Hydroxypropylmethylcellulose	Non-ionic	+
Poly (vinyl alcohol)	Non-ionic	+
Poly (vinyl pyrrolidone)	Non-ionic	+

Mucoadhesive strength +++ excellent; ++ good; + poor/absent

References

Abdelkader H, Ismail S, Kamal A, Alany RG (2011) Design and evaluation of controlled release niosomes and discomes for naltrexone hydrochloride ocular delivery. J Pharm Sci 100: 1833–1846

Abrishami M, Zarei-Ganavati S, Soroush D, Rouhbakhsh M, Jaafari MR, Malaekeh Nikouei B (2009) Preparation, characterization, and *in vivo* evaluation of nanoliposomes-encapsulated bevacizumab (avastin) for intravitreal adminis-tration. Retina 29:699–703

Abuzaid SS, El-Ghamry HA, Hammad M (2003) Liposomes as ocular drug delivery system for atenolol. Egypt J Pharm Sci 44:227–245

Addo RT, Siddig A, Patel NJ, Siwale R, Akande J, Uddin AU, D'Souza MJ (2010) Formulation, characterization, and testing of tetracaine hydrochloride-loaded albumin-chitosan microparticles for ocular drug delivery. J Microencapsul 27(2):95–104

Addo RT, Yeboah KG, Siwale RC, Siddig A, Jones A, Ubale RV, Akande J, Nettey H, Patel NJ, Addo E, D'Souza MJ (2015) Formulation and characterization of atropine sulfate in albumin-chitosan microparticles for in vivo ocular drug delivery. J Pharm Sci 104(5): 1677–1690

Aggarwal D, Kaur IP (2005) Improved pharmacodynamics of timolol maleate from mucoadhesive niosomal ophthalmic drug delivery system. Int J Pharm 16:155–159

Aggarwal D, Pal D, Mitra AK, Kaur IP (2007) Study of the extent of ocular absorption of acetazolamide from a developed niosomal formulation, by micro dialysis sampling of aqueous humor. Int J Pharm 29:21–26

Agnihotri SM, Vavia PR (2009) Diclofenac-loaded biopolymeric nanosuspensions for ophthalmic application. Nanomed Nanotech Boil Med 5:90–95

Ahmed I, Patton TF (1985) Importance of the noncorneal absorption route in topical ophthalmic drug delivery. Invest Ophthalmol Vis Sci 26(4):584–587

Ahmed I, Patton TF (1987) Disposition of timolol and inulin in the rabbit eye following corneal versus non-corneal absorption. Int J Pharm 38:9–21

Ahmed S Guinedi, Nahed DM, Samar M, Rania MH (2005) Preparation and evaluation of reverse-phase evaporation and multilamellar niosomes as ophthalmic carriers of acetazolamide. Int J Pharm 306:71–82

Akhter S, Talegaonkar S, Khan ZI, Jain GK, Khar RK, Ahmad FJ (2011) Assessment of ocular pharmacokinetics and safety of ganciclovir loaded nanoformulation. Biomed Nanotechnol 7:144–145

Aksungur P, Demirbilek M, Denkbas EB, Vandervoort J, Ludwig A, Unlu N (2011) Development & characterization of cyclosporine A loaded nanoparticles for ocular drug delivery: cellular toxicity, uptake and kinetic studies. J Control Rel 151:286–294

Ameeduzzafar Ali J, Bhatnagar A, Kumar N, Ali A (2014) Chitosan nanoparticles amplify the ocular hypotensive effect of cateolol in rabbits. Int J Biol Macromol 65:479–491

Ammar HO, Salama HA, Ghorab M, Mahmoud AA (2009) Nanoemulsion as a potential ophthalmic delivery system for dorzolamide hydrochloride. AAPS PharmSciTech 10(3):808

Araujo J, Gonzalez E, Egea MA, Garcia ML, Souto EB (2009) Nanomedicines for ocular NSAIDs: safety on drug delivery. Nanomedicine 5:394–401

Arunothayanun P, Bernard MS, Craig DQ, Uchegbu IF, Florence AT (2000) The effect of processing variables on the physical characteristics of non-ionic surfactant vesicles (niosomes) formed from hexadecyl diglycerol ether. Int J Pharm 201:7–14

Badawi AA, El-Laithy HM, El Qidra RK, El Mofty H, El dally M (2008) Chitosan based nanocarriers for indomethacin ocular delivery. Arch Pharm Res 31(8):1040-1049

Bai S, Thomas C, Rawat A, Ahsan F (2006) Recent progress in dendrimer-based nanocarriers. Crit Rev Ther Drug Carrier Syst 23:437–495

Balakrishnan P, Shanmugam S, Lee WS, Lee WM (2009) Formulation and *in vitro* assessment of minoxidil niosomes for enhanced skin delivery. Int J Pharm 377:1–8

Bangham D, Standish MM, Watkins JC (1965) Diffusion of univalent ions across the lamellae of swollen phospholipids. J Mol Biol 13:238–252

Barbault-Foucher S, Gref R, Russo P, Guechot J, Bochot A (2002) Design of poly–caprolactone nanospheres coated with bioadhesive hyaluronic acid for ocular delivery. J Control Release 83(3):365–375

Başaran, E, Demirel, M, Sırmagül, B, Yazan, Y (2011) Polymeric cyclosporine-A nanoparticles for ocular application. J Biomed Nanotechnol 7(5):714–723

Başaran E, Yenilmez E, Berkman MS, Büyükköroğlu G, Yazan Y (2014) Chitosan nanoparticles for ocular delivery of cyclosporine A. J Microencapsul 31(1):49–57

Basha M, Hosam , El-Alim A, Shamma RN, Awad GEA (2013) Design and optimization of surfactant based nanovesicles for ocular delivery of clotrimazole

Baspinar Y, Bertelmann E, Pleyer U, Buech G, Siebenbrodt I, Borchert HH (2008) Corneal permeation studies of everolimus microemulsion. J Ocul Pharmacol Ther 24(4):399–402

Biswal S, Murthy PN, Sahu J, Sahoo P, Amir F (2008) Vesicles of non-ionic surfactants (niosomes) and drug delivery potential. Int J Pharm Sci Nanotechnol 1:1–8

Bochot A, Fattal E, Grossiord JL, Puisieux F, Couvreur P (1998a) Characterization of a new drug delivery system based on dispersion of liposomes in a thermosensitive gel. Int J Pharm 162:119–127

Bochot A, Gulik FA, Couarraze G, Couvreur P (1998b) Liposomes dispersed within a thermosensitive gel: a new dosage form for ocular delivery of oligonucleotides. Pharm Res 15:1364–1369

Bourlais CL, Acar L, Zia H, Sado PA, Needham T, Leverge R (1998) Ophthalmic drug delivery systems: recent advances. Prog Retin Eye Res 17:33–58

Calvo P, Remunan-Lopez C, Vila-Jato JL, Alonso MJ (1997) Novel hydrophilic chitosan-polyethylene oxide nanoparticles as protein carriers. J Appl Polym Sci 63:125–132

Carafa M, Santucci E, Lucania G (2002) Lidocaine loaded non ionic surfactant vesicles: characterization and in vitro permeation studies. Int J Pharm 231:21–32

Carol L, Keita AV (2010) Translocation of Crohn's disease Escherichia coli across M cells: contrasting effects of soluble plant fibers and emulsifiers. Gut 59:1331–1339

Chen Y, Lu Y, Zhong Y, Wang Q, Wu W, Gao S (2012) Ocular delivery of cyclosporine A based on glyceryl monooleate/poloxamer 407 liquid crystallinenanoparticles: preparation, characterization, in vitro corneal penetration and ocular irritation. J Drug Target 20(10):856–863

Chetoni P, Rossi S, Burgalassi S, Monti D, Mariotti S, Saettone MF (2004) Comparison of liposome-encapsulated acyclovir with acyclovir ointment: ocular pharmacokinetics in rabbits. J Ocul Pharmacol Ther 20:169–177

Cruysberg LP, Nuijts RM, Geroski DH, Koole LH, Hendrikse F, Edelhauser HF (2002) In vitro human scleral permeability of fluorescein, dexamethasone-fluorescein, methotrexate-fluorescein and rhodamine 6G and the use of a coated coil as a new drug delivery system. J Ocul Pharmacol Ther 18:559–569

Dai Y, Zhou R, Liu L, Lu Y, Qi J, Wu W (2013) Liposomes containing bile salts as novel ocular delivery systems for tacrolimus (FK506): in vitro characterization and improved corneal permeation. Int J Nanomed 8:1921–1933

Danion A, Arsenault I, Vermette P (2007) Antibacterial activity of contact lenses bearing surface-immobilized layers of intact liposomes loaded with levofloxacine. J Pharm Sci 96:2350–2363

De Campos AM, Sanchez A, Gref R, Calvo P, Alonso MJ (2003) The effect of a PEG versus a chitosan coating on the interaction of drug colloidal carriers with the ocular mucosa. Eur J Pharm Sci 20:73–81

De Campos AM, Diebold Y, Carvaiho ELS, Sanchez A, Alonso MJ (2004) Chitosan nanoparticles as new ocular drug delivery system: in vitro stability, in vivo fate, and cellular toxicity. Pharm Res 21:803–810

Dhubhghaill SN, Humphries MM, Kenna PF (2012) Further development of barrier modulation as a technique for systemic ocular drug delivery. Adv Exp Med Biol 723:155–159

Ding S (1998) Recent developments in ophthalmic drug delivery. Pharm Sci Technol Today 1:328–335

Ding X, Alani WG, Robinson JR (2005) Extended-release and targeted drug delivery systems. In: Troy DB (ed) Remington: the science and practice of pharmacy. Lippincott Williams and Wilkins, Philadelphia, PA, USA

Djekic L, Ibric S, Primorac M (2008) The application of artificial neural networks in the prediction of microemulsion phase boundaries in PEG-8 caprylic/capric glycerides based systems. Int J Pharm 361:41–46

Du Toit LC, Govender T, Carmichael T, Kumar P, Choonara YE, Pillay V (2013) Design of an anti-inflammatory composite nanosystem and evaluation of its potential for ocular drug delivery. J Pharm Sci 102(8):2780–2805

Duchfine D, Touchard F, Peppas NA (1988) Pharmaceutical and medical aspects of bioadhesive systems of drug administration. Drug Dev Ind Pharm 14:283–318

Elbayoumi TA, Torchilin VP (2010) Current trends in liposome research. Methods Mol Biol 605:1–27

Felt O, Furrer P, Mayer JM, Plazonnet B, Buri P, Gurny R (1999) Topical use of chitosan in ophthalmology: tolerance assessment and evaluation of pre-corneal retention Int J Pharm 180:185–193

Fessi H, Puisieux F, Devissaguet JP, Ammoury N, Benitas S (1989) Nanocapsule formation by interfacial polymer deposition following solvent displacement. Int J Pharm 55:R1–R4

Fialho SL (2004) New vehicle based on a microemulsion for topical ocular administration of dexamethasone. Clin Exp Ophthalmol 32:626–632

Fischbarg J (2006) The corneal endothelium. In: Fischbarg J (ed) The Biology of Eye. Academic Press, New York, NY, USA, pp 113–125

Fitzgerald P, Wilson C (1994) Polymeric systems for ophthalmic drug delivery. In: Severian D (ed) Polymeric systems for ophthalmic drug delivery. Marcel Dekker, New York. 373–398

Gajbhiye V, Palanirajan VK, Tekade RK, Jain NK (2009) Dendrimers as therapeutic agents: a systematic review. J Pharm Pharmacol 61:989–1003

Garty N, Lusky M (1994) Pilocarpine in submicron emulsion formulation for treatment of ocular hypertension: a phase II clinical trial. Invest Ophthalmol Vis Sci 35:2175–2185

Gasco MR, Gallarate M, Trotta M, Bauchiero L, Gremmo E and Chiappero O (1989) Microemulsions as topical delivery vehicles: Ocular administration of timolol. J Pharm Biomed Analysis 7(4):433–439

Gausas RE, Gonnering RS, Lemke BN, Dortzbach RK, Sherman DD (1999) Identification of human orbital lymphatics. Ophthal Plast Reconstr Surg 15:252–259

Gipson IK, Argueso P (2003) Role of mucin in the function of the corneal and conjunctival epithelia. Int Rev Cytol 231:1–49

Govender S, Pillay V, Chetty DJ, Essack SY, Dangor CM, Govender T (2005) Optimization and characterization of bioadhesive controlled release tetracycline microspheres. Int J Pharm 306:24–40

Greaves JL, Wilson CG, Birmingham AT (1993) Assessment of the precorneal residence of an ophthalmic ointment in healthy subjects. Br J Clin Pharmacol 35:188–192

Guinedi AS, Mortada ND, Mansour S, Hathout RM (2005) Preparation and evaluation of reverse-phase evaporation and multilamellar niosomes as ophthalmic carriers of acetazolamide. Int J Pharm 306:71–82

Gupta S, Moulik SP (2008) Biocompatible Microemulsions and their prospective uses in drug delivery. J Pharm Sci 97:22–45

Gupta H, Aqil M, Khar RK, Ali A, Bhatnagar A, Mittal G (2010) Sparfloxacin-loaded PLGA nanoparticles for sustained ocular drug delivery. Nanomed Nanotech Biol Med 6:324–333

Gupta H, Aqil M, Khar RK, Ali A, Bhatnagar A, Mittal G (2011) Biodegradable levofloxacine nanoparticles for sustained ocular drug delivery. J Drug Target 19(6):409–417

Hanrahan F, Campbell M, Nguyen AT, Suzuki M, Kiang AS, Tam LC, Gobbo OL, Dhubhghaill SN, Humphries MM, Kenna PF (2012) Further development of barrier modulation as a technique for systemic ocular drug delivery. Adv Exp Med Biol 723:155–159

Hait SK and Moulik SP (2002) Gemini surfactants: A distinct class of self-assembling molecules. Curr Sci 82:1101–1111

Hamdy A, Sayed I, Amal K, Raid GA (2011) Design and evaluation of controlled-release niosomes and discomes for naltrexone hydrochloride ocular delivery. J Pharm Sci 100(5): 1833–1846

Hathout RM, Mansour S, Mortada ND (2007) Liposomes as an ocular delivery system for acetazolamide: in vitro and in vivo studies. AAPS Pharm Sci Tech 8:1–12

Henriksen I, Green KL, Smart JD, Smistad G, Karlsen J (1996) Bioadhesion of hydrated chitosans: an in vitro and in vivo study. Int J Pharm 145:231–240

Her Y, Lim JW, Han SH (2013) Dry eye and tear film functions in patients with psoriasis. Jpn J Ophthalmol 57(4):341–346

Higashiyama M, Tajika T, Inada K, Ohtori A (2006) Improvement of the ocular bioavailability of carteolol by ion pair. J Ocular Pharmacol Ther 22:333–339

Hill JM, O'Callaghan RJ, Hobden JA, Kaufman E (1993) Corneal collagen shields for ocular drug delivery. Ophthalmic drug delivery systems. Marcel Dekker, New York, pp 261–275

Hirano S, Seino H, Akiyama I, Nonaka I (1990) Chitosan: a biocompatible material for oral and intravenous administration. In: Gebelein CG, Dunn RL (eds) Progress in biomedical polymers. Plenum Press, New York, pp 283–289

Hitzenberger CK, Baumgartner A, Drexler W et al (1994) Interferometric measurement of corneal thickness with micrometer precision. Am J Ophthalmol 118:468–476

Honda M, Asai T, Oku N, Araki Y, Tanaka M, Ebihara N (2013) Liposomes and nanotechnology in drug development: focus on ocular targets. Int J Nanomed 8:495–503

Hosoya K, Vincent HL, Kim LKJ (2005) Roles of the conjunctiva in ocular drug delivery: a review of conjunctival transport mechanisms and their regulation. Eur J Biopharm 60:227–240

Ikeda I (2003) Synthesis of gemini and elated substances. In: Zana R, Xia J (eds) Gemini surfactants synthesis, interfacial and solution phase-behavior, and applications, vol 1. Marcel Dekker, New York, pp 26–30

Irache JM, Merodio M, Arnedo A, Camapanero MA, Mirshahi M, Espuelas S (2005) Albumin nanoparticles for the intravitreal delivery of anticytomegaloviral drugs. Mini Rev Med Chem 5:293–305

Jain NK, Gupta U (2008) Application of dendrimer-drug complexation in the enhancement of drug solubility and bioavailability. Expert Opin Drug Discov 4:1035–1051

Jain K, Kumar RS, Sood S, Dhyanandhan G (2013) Betaxolol hydrochloride loaded chitosan nanoparticles for ocular delivery and their anti-glaucoma efficacy. Curr Drug Deliv 10:493–499

Jain S, Thompson JR, Foot B, Tatham A, Eke T (2014) Severe intraocular pressure rise following intravitreal triamcinolone: a national survey to estimate incidence and describe case profiles. Eye (Lond) 28(4):399–401. doi:10.1038/eye.2013.306

Javadzadeh Y, Ahadi F, Davaran S, Mohammadi G, Sabzevari A, Adibkia K (2010) Preparation and physicochemical characterization of naproxen-PLGA nanoparticles. Colloids Surf B Biointerfaces 81:498–502

Jesorka A, Orwar O (2008) Liposomes: technologies and analytical applications. Ann Rev Anal Chem 1:801–832

Jóhannesson G, Moya-Ortega MD, Asgrímsdottir GM, Agnarsson BA, Lund SH, Loftsson T, Stefansson E (2014) Dorzolamide cyclodextrin nanoparticle suspension eye drops and Trusopt in rabbit. J Ocul Pharmacol Ther 30(6):464–467

Jtirvinen K, Tomi J, Arto Urttia S (1995) Ocular absorption following topical delivery. Adv Drug Deliv Rev 16:3–19

Jwala J, Boddu SHS, Shah S, Sirimulla S, Pal D, Mitra AK (2011) Ocular sustained release nanoparticles containing stereoisomeric dipeptide prodrugs of acyclovir. J Ocul Pharmacol Ther 27(2):163–172

Kalam MA (2016) The potential application of hyaluronic acid coated chitosan nanoparticles in ocular delivery of dexamethasone. Int J Biol Macromol 6(89):559–568
Kambhampati SP, Kannan RM (2013) Dendrimer nanoparticles for ocular drug delivery. J Ocul Pharmacol Ther 29(2):151–165
Kapadia R, Khambete H, Katara R (2009) Novel approach for ocular delivery of acyclovir via niosomes entrapped in situ hydrogel system. J Pharm Res 2:745–751
Karim KM, Mandal AS, Biswas N, Guha A, Chatterjee S, Behera M, Kuotsu K (2012) Niosome: a future of targeted drug delivery systems. J Adv Pharm Tech Res 1:374–380
Katiyar S, Pandit J, Mondal RS, Mishra AK, Chuttani K, Aqil M, Ali A, Sultana Y (2014) In situ gelling dorzolamide loaded chitosan nanoparticles for the treatment of glaucoma. Carbohydr Polym 15(102):117–124
Kaur IP, Kanwar M (2002) Ocular preparations: the formulation approach. Drug Dev Ind Pharm 28:473–493
Kaur IP, Garg A, Singla AK, Aggarwal D (2004) Vesicular systems in ocular drug delivery: an overview. Int J Pharm 269:1–14
Kaur IP, Aggarwal D, Singh H, Kakkar S (2010) Improved ocular absorption kinetics of timolol maleate loaded into a bioadhesive niosomal delivery system. Graefes Arch Clin Exp Ophthalmol 248:1467–1472
Kaur H, Ahuja M, Kumar S, Dilbaghi N (2012) Carboxymethyl tamarind kernel polysaccharide nanoparticles for ophthalmic drug delivery. Int J Biol Macromol 50:833–839
Kesavan K, Kant S, Singh PN, Pandit JK (2013) Mucoadhesive chitosan-coated cationic microemulsion of dexamethasone for ocular delivery: in vitro and in vivo evaluation. Curr Eye Res 38:342–352
Khan A, Sharma PK, Visht S, Malviya R (2011) Niosomes as colloidal drug delivery system: a review. J Chronotherapy Drug Deliv 2:15–21
Khazaeli P, Pardakhty A, Shoorabi H (2007) Caffeine-loaded niosomes: characterization and in vitro release studies. Drug Deliv 14:447–452
King-Smith PE, Fink BA, Fogt N et al (2000) The thickness of the human precorneal tear film: evidence from reflection spectra. Invest Ophthalmol Visual Sci 41:3348–3359
Klibanov AL, Maruyama K, Torchilin VP, Huang L (1990) Amphipathic polyethyleneglycols effectively prolong the circulation time of liposomes. FEBS Lett 268:235–237
Kumari A, Yadav SK, Yadav SC (2010) Biodegradable polymeric nanoparticles based drug delivery systems. Colloids Surf A 72(1):1–18
Kumbhar D, Wavikar P, Vavia P (2013) Niosomal gel of lornoxicam for topical delivery: in vitro assessment and pharmacodynamic activity. AAPS Pharm Sci Tech 14(3):1072–1082
Lang JC (1995) Ocular drug delivery conventional ocular formulations. Advanced Drug Delivery Reviews 16:39–43
Lawrence MJ, Chauhan S, Lawrence SM, Barlow DJ (1996) The formation, characterization and stability of non-ionic surfactant vesicles. STP Pharm Sci 1:49–60
Lee SJ, He W, Robinson SB, Robinson MR, Csaky KG, Kim H (2010) Evaluation of clearance mechanisms with trans-scleral drug delivery. Invest Ophthalmol Vis Sci 51:5205–5212
Li J, Wu L, Wu W, Wang B, Wang Z, Xin H, Xu Q (2013) A potential carrier based on liquid crystal nanoparticles for ophthalmic delivery of pilocarpine nitrate. Int J Pharm 15 455(1–2):75–84
Lin J, Wu H, Wang Y, Lin J, Chen Q, Zhu X (2016) Preparation and ocular pharmacokinetics of hyaluronan acid-modified mucoadhesive liposomes. Drug Deliv 23(4):1144–1151
Liu R, Wang S, Fang S, Wang J, Chen J, Huang X, He X, Liu C (2016) Liquid crystalline nanoparticles as an ophthalmic delivery system for tetrandrine: development, characterization, and in vitro and in vivo. Nanoscale Res Lett 11(1):254. doi:10.1186/s11671-016-1471-0 Cited 2016 May 17
Ludwig A (2005) The use of mucoadhesive polymers in ocular drug delivery. Adv Drug Delivery Rev 57:1595–1639
Ludwig A, Ooteghem VM (1992) Influence of, viscoslysers on the residence of ophthalmic solutions evaluated by slip lamp fluorophotometry. STP Pharm Sci 2(1):81–87

Lv FF, Zheng LQ, Tung CH (2005) Phase behavior of the microemulsions and the stability of the chloramphenicol in the microemulsion-based ocular drug delivery system. Int J Pharm 301(1–2):237–246
Ma SW, Gan Y, Gan L, Zhu CL, Zhu JB (2008) Preparation and in vitro corneal retention behaviour of novel cationic microemulsion/in situ gel system. Yao Xue Xue Bao 43:749–755
Mahale NB, Thakkar PD, Mali RG, Walunj DR, Chaudhari SR (2012) Niosomes: novel sustained release nonionic stable vesicular systems—an overview. Adv Colloid Interface Sci 15:46–54
Mainardes RM, Urban MCC, Cinto PO (2005) Colloidal carriers for ophthalmic drug delivery. Curr Drug Targets 6:363–371
Manosroi A, Wongtrakul P, Manosroi J, Sakai H, Sugawara F, Yuasa N (2003) Characterization of vesicles prepared with various nonionic surfactants mixed with cholesterol. Colloids Surf B Biointerfaces 30:129–138
Marfurt CF, Kingsley RE, Echtenkamp SE (1989) Sensory and sympathetic innervation of the mammalian cornea. A retrograde tracing study. Invest Ophthalmol Vis Sci 30:461–472
Maurice DM, Mishima S (1984) Pharmacology of the eye. Handb Exp Pharmacol 69:109–116
McDonald TO, Shadduck JA (1977) Eye irritation. Adv Mod Toxicol 4:139–191
Meisner D, Mezei M (1995) Liposome ocular delivery systems. Adv Drug Deliv Rev 16:75–93
Meng J, Sturgis TF, Youan BC (2011) Engineering tenofovir loaded chitosan nanoparticles to maximize microbicide mucoadhesion. Eur J Pharm Sci 44:57–67
Merodio M, Irache JM, Valamanesh F, Mirshahi M (2002) Ocular disposition and tolerance of ganciclovir-loaded albumin nanoparticles after intravitreal injection in rats. Biomaterials 23:1587–1594
Meseguer G, Buri P, Plazonnet B, Rozier A, Gurny R (1996) Gamma scintigraphic comparison of eyedrops containing pilocarpine in healthy volunteers. J Ocul Pharmacol Ther 12:481–488
Miao H, Wu B, Tao Y, Li X. (2013). Diffusion of macromolecules through sclera. Acta Ophthamologica 91(1):e1–e6
Mintzer MA, Grinstaff MW (2011) Biomedical applications of dendrimers: a tutorial. Chem Soc Rev 40(1):173–190
Mishra GP, Bagui M, Tamboli V, Mitra AK (2011) Recent applications of liposomes in ophthalmic drug delivery. J Drug Deliv 14:205–2018
Mohammed N, Rejinold NS, Mangalathillam S, Biswas R, Nair SV, Jayakumar R (2013) Fluconazole loaded chitin nanogels as a topical ocular drug delivery agent for corneal fungal infections. J Biomed Nanotechnol 9(9):1521–1531
Monem AS, Ali FM, Ismail MW (2000) Prolonged effect of liposomes encapsulating pilocarpine HCl in normal and glaucomatous rabbits. Int J Pharmaceutics 198:29–38
Montes-Mico R, Cervino A, Ferrer-Blasco T, Garcia-Lazaro S, Madrid-Costa D (2010) The tear film and the optical quality of the eye. Ocul-Surf 8:185–192
Motwani SK, Chopra S, Talegaonkar S, Kohli K, Ahmad FJ, Khar RK (2008) Chitosan–sodium alginate nanoparticles as submicroscopic reservoirs for ocular Delivery: formulation, optimization and in-vitro characterization. Eur J Pharm Biopharm 68:513–525
Mujoriya RZ, Bodla RB (2011) Niosomes—challenge in preparation for pharmaceutical scientist. Int J App Pharm 3:11–15
Muller LJ, Marfurt CF, Kruse F, Tervo TM (2003) Corneal nerves: structure, contents and function. Exp Eye Res 76:521–542
Myles ME, Neumann DM, Hill JM (2005) Recent progress in ocular drug delivery for posterior segment disease: emphasis on transscleral iontophoresis. Adv Drug Del Rev 57:2063–2079
Nagarwal RC, Kumar R, Pandit JK (2012) Chitosan coated sodium alginate-chitosan nanoparticles loaded with 5-FU for ocular delivery: *in vitro* characterization and *in vivo* study in rabbit eye. Eur J Pharm Sci 47(4):678–685
Nagyova B, Tiffany JM (1999) Components responsible for the surface tension of human tears. Curr Eye Res 19:4–11
Newell FW (1993) Ophthalmology: principles and concepts. 7th edn. Mosby Year Book, St. Louis

Newkome GR, Shreiner CD (2008) Amidoamine, polypropylenimine, and related dendrimers and dendrons possessing different 1 → 2 branching motifs: An overview of the divergent procedures. Polymer 49:1–173

Norley SG, Huang L, Rouse BT (1986) Targeting of drug loaded immunoliposomes to herpes simplex virus infected corneal cells: an effective means of inhibiting virus replication in vitro. J Immunol 136(2):681–685

Omidi, Y., Barar, J., Hamzeiy, H., 2012. Nanomedicines Impacts in Ocular Delivery and Targeting. Nanotechnol Health Care, 43–106

Pandit JC, Nagyova B, Bron AJ, Tiffany JM (1999) Physical properties of stimulated and unstimulated tears. Exp Eye Res 68:247–253

Pardakhti A, Moshefi MH, Moteshafi H (2007) Preparation of niosomes containing chloramphenicol sodium succinate and evaluation of their physicochemical and antimicrobial properties. Pharm Sci Spr 1:11–21

Parra A, Mallandrich M, Clares B, Egea MA, Espina M, García ML, Calpena AC (2015) Design and elaboration of freeze-dried PLGA nanoparticles for the transcorneal permeation of carprofen:Ocular anti-inflammatory applications. Colloids Surf B Biointerfaces 136:935–943

Pathak MK, Chhabra G, Pathak K (2013) Design and development of a novel pH triggered nanoemulsified in-situ ophthalmic gel of fluconazole: *ex-vivo* transcorneal permeation, corneal toxicity and irritation testing. Drug Dev Ind Pharm 39:780–790

Pescina S, Santi P, Ferrari G, Nicoli S (2011) Trans-scleral delivery of macromolecules. Ther Deliv 2:1331–1349

Pignatello R, Bucolo C, Spedalieri G, Maltese A, Puglisi G (2002) Flurbiprofen-loaded acrylate polymer nanosuspensions for ophthalmic application. Biomaterials 23:3247–3255

Ponchel G, Touchard F, Duchfine D, Peppas NA (1987) Bioadhesive analysis of controlled-release systems. I. Fracture and interpenetration analysis in poly (acrylic acid)-containing systems. J Control Release 5:129–141

Qi HP, Gao XC, Zhang LQ, Wei SQ, Bi S, Yang ZC, Cui H (2013) *In vitro* evaluation of enhancing effect of borneol on transcorneal permeation of compounds with different hydrophilicities and molecular sizes. Eur J Pharmacol 705:20–25

Rathore KS, Nema PK (2009) An insight into ophthalmic drug delivery system. Int J Pharm Sci Drug Res 1:1–5

Raviola G (1983) Conjunctival and episcleral blood vessels are permeable to blood-borne horseradish peroxidase. Invest Ophthalmol Vis Sci 24:725–736

Raymzond CR, Paul JS, Sian CO (2006) Polyoxyethylenealkyl ethers. Handbook of pharmaceutical excipients, 5th edn. Pharmaceutical Press, London, pp 564–571

Robinson JC (1993) Ocular anatomy and physiology relevant to ocular drug delivery. In: Mitra AK (ed) Ophthalmic drug delivery systems. Marcel Dekker, New York, pp 29–57

Rojanasakul Y, Wang LY, Bhat M (1992) The transport barrier of epithelia: a comparative study on membrane permeability and charge selectivity in the rabbit. Pharm Res 9:1029–1934

Rooijen VN, Nieuwmegen VR (1980) Liposomes in immunology: multilamellar phosphatidylcholine liposomes as a simple, biodegradable and harmless adjuvant without any immunogenic activity of its own. Immunol Commun 9:243–256

Round A, Berry M, McMaster T, Stoll S, Gowers D, Corfield A, Miles M (2002) Heterogeneity and persistence length in human ocular mucins. Biophys J 83:1661–1670

Sabzevari A, Adibkia K, Hashemi H, De-Geest BG, Mohsenzadeh N, Atyabi F, Ghahremani MH, Khoshayand MR, Dinarvand R (2013) Improved anti-inflammatory effects in rabbit eye model using biodegradable poly beta-amino ester nanoparticles of triamcinolone acetonide. Invest Ophthalmol Vis Sci 54:5520–5526

Sahoo SK, Labhasetwar V (2003) Nanotech approaches to drug delivery and imaging. Drug Discov Today 8:1112–1120

Sahoo SK, Dilnawaz F, Krishnakumar S (2008) Nanotechnology in ocular drug delivery. Drug Discov Today 13:144–151

Salama AH, Mahmoud AA, Kamel R (2015) A novel method for preparing surface-modified fluocinolone acetonide loaded PLGA nanoparticles for ocular use: in vitro and in vivo evaluations. AAPS Pharm Sci Tech, 1–14

Samad A, Alam MI, Saxena K (2009) Dendrimers: a class of polymers in the nanotechnology for the delivery of active pharmaceuticals. Curr Pharm 15(25):2958–2969

Sasaki H, Karasawa K, Hironaka K, Tahara K, Tozuka Y, Takeuchi H (2013) Retinal drug delivery using eyedrop preparations of poly-L-lysine-modifiedliposomes. Eur J Biopharm 83:364–369

Schipper NG, Olsson S, Hoogstraate JA, deBoer AG, Varum KM, Artursson P (1997) Chitosans as absorption enhancers for poorly absorbable drugs 2: mechanism of absorption enhancement. Pharm Res 14:923–929

Schoenwald RD (1990) Ocular drug delivery: pharmacokinetic considerations. Clin Pharmacokinet 18:255–269

Schoenwald RD, Ward RL, Desantis LM, Roehrs RE (1978) Influence of high viscosity vehicles on miotic effect of pilocarpine. J Pharm Sci 67:1280–1283

Schoenwald RD, Vidvauns S, Wurster DE, Barfknecht CF (1998) The role of tear proteins in tear film stability in the dry eye patient and in the rabbit. In: Sullivan DA, Dartt DA and Menerary MA (eds) Lacrimal gland, Tear Film and Dry Eye Sydromes. 2. Plenum Press: New York, pp 391–400

Sechoy O, Tissiiee G, Sebastian C, Maurin F, Driot JY, Trinquand C (2000) A new long acting ophthalmic formulation of Carteolol containing alginic acid. Int J Pharm 207:109–116

Shahiwala A, Misra A (2002) Studies in topical application of niosomally entrapped nimesulide. J Pharm Sci 5:220–225

Shan W, Liu H, Shi J, Yang L, Hu N (2008) Self-assembly of electroactive layer-by-layer films of heme proteins with anionic surfactant dihexadecyl phosphate. Biophys Chem 134:101–109

Shen Y, Tu J (2007) Preparation and ocular pharmacokinetics of ganciclovir liposomes. AAPS Pharm Sci Tech 9:371–377

Singh D (2003) Conjunctival lymphatic system. J Cataract Refract Surg 29:632–633

Singh V, Ahmad R, Heming T (2011) The challenges of ophthalmic drug delivery: a review. Int J Drug Discov 3:56–62

Slovin EM, Robinson JR (1993) Bioadhesive in ocular drug delivery. In: Edman P (ed) Biopharmaceutics of ocular drug delivery. CRC Press, Boca Raton, Florida, pp 145–157

Soukharev AR, Wojciechowska J (2005) Microemulsions as potential ocular drug delivery systems: phase diagrams and physical depending on ingredients. Acta Pol Pharm Drug Res 62:465–471

Stahl U, Willcox M, Stapleton F (2012) Osmolality and tear film dynamics. Clin Exp Optom 95(1):3–11

Sugar HS, Riazi A, Schaffner R (1957) The bulbar conjunctival lymphatics and their clinical significance. Trans Am Acad Ophthalmol Otolaryngol 61:212–223

Sun KX, Wang AP, Huang LJ, Liang RC, Liu K (2006) [Preparation of diclofenac sodium liposomes and its ocular pharmacokinetics]. Yao Xue Xue Bao. 41(11):1094–1098

Szczesna DH, Jaronski J, Kasprzak HT, Stenevi U (2006) Interferometric measurements of dynamic changes of tear film. J Biomed Opt 11(3):340–352

Szczesna DH, Kasprzak HT, Jaronski J, Rydz A, Stenevi U (2007) An interferometric method for the dynamic evaluation of the tear film. Acta Ophthalmol Scand 85:202–208

Szczesna-Iskander DH, Iskander DR (2012) Future directions in non-invasive measurements of tear film surface kinetics. Optom Vis Sci 89(5):749–759

Szulc J, Woyczikowski B, De Laval W (1988) Effect of pilocarpine hydrochloride liposomes on the intraocular pressure of the rabbit eye pupil. Farm Pol 44:462–465

Takeuchi H, Yamamoto H, Niwa T, Hino T, Kawashima Y (1996) Enteral absorption of insulin in rats from mucoadhesive chitosan coated liposomes. Pharm Res 13:896–901

Tayel SA, El-Nabarawi MA, Tadros MI, Abd-Elsalam WH (2013) Positively charged polymeric nanoparticle reservoirs of terbinafine hydrochloride: preclinical implications for controlled drug delivery in the aqueous humor of rabbits. AAPS Pharm Sci Tech 14:782–793

Tiffany JM (2003) Tears in health and disease. Eye 17:1–4
Tissie G, Sebastian C, Elena PP, Driot JY, Trinquand C (2002) Alginic acid effect on carteolol ocular pharmacokinetic in the pigmented rabbit. J Ocul Pharmacol Ther 18:65–73
Van Ooteghem M (1995) Preparations ophtalmiques. In: Galenica (ed) Technique and documentation. Lavoisier, Paris
Vandamme TF, Brobeck L (2005) Poly (amidoamine) dendrimers as ophthalmic vehicles for ocular delivery of pilocarpine nitrate and tropicamide. J Control Release 102(1):23–38
Vandervoort J, Ludwig A (2004) Preparation and evaluation of drug-loaded gelatin nanoparticles for topical ophthalmic use. Eur J Biopharm 57:251–261
Vyas, S.P., Khar, R.K., 2008. Ocular drug delivery, in controlled drug delivery. 384–385
Wadhwa S, Paliwal R, Paliwal SR, Vyas SP (2009) Nanocarriers in ocular drug delivery: an update review. Curr Pharm Des 15:2724–2750
Wadhwa S, Paliwal R, Paliwal SR, Vyas SP (2010) Hyaluronic acid modified chitosan nanoparticles for effective management of glaucoma: development, characterization, and evaluation. J Drug Target 18(4):292–302
Warsi MH, Anwar M, Garg V, Jain GK, Talegaonkar S, Ahmad FJ, Khar RK (2014) Dorzolamide-loaded PLGA/vitamin E TPGS nanoparticles for glaucoma therapy: pharmacoscintigraphy study and evaluation of extended ocular hypotensive effect in rabbits. Colloids Surf B Biointerfaces 122:423–431
Wen H, Hao J, Li SK (2013) Characterization of human sclera barrier properties for transscleral delivery of bevacizumab and ranibizumab. J Pharm Sci 102:892–903
Wenger Y, Schneider RJ, Reddy GR, Kopelman R, Jolliet O, Philbert MA (2011) Tissue distribution and pharmacokinetics of stable polyacrylamide nanoparticles following intravenous injection in the rat. Toxicol Appl Pharmacol 251:181–190
Wilson CG (2004) Topical drug delivery in the eye. Exp Eye Res 78(3):737–743
Wolff E (1946) The muco-cutaneous junction of the lid margin and the distribution of the tear fluid. Trans Ophthamol Soc UK 66:291–308
Wolff E (1954) The Anatomy of the Eye and Orbit, fourth ed. H.K Lewis and Co, London. 49
Worakul N, Robinson JR (1997) Ocular pharmacokinetics-pharmacodynamics. Eur J Pharm Biopharm 44(1627):71–83
Yan Y, Ting L, Huang J (2009) Recent advances in bolaamphiphiles and oppositely charged conventional surfactants. J Colloid Interface Sci 337:1–10
Yang H, Kao WJ (2006) Dendrimers for pharmaceutical and biomedical applications. J Biomater Sci Polym Ed 17:13–19
Zeng W, Li Q, Wan T, Liu C, Pan W, Wu Z, Zhang G, Pan J, Qin M, Lin Y, Wu C, Xu Y (2016) Hyaluronic acid-coated niosomes facilitate tacrolimus ocular delivery: mucoadhesion, precorneal retention, aqueous humor pharmacokinetics, and transcorneal permeability. Colloids Surf B Biointerfaces 1(141):28–35
Zhou W, Wang Y, Jian J, Song S (2013) Self-aggregated nanoparticles based on amphiphilic poly (lactic acid)-grafted-chitosan copolymer for ocular delivery of amphotericin B. Int J Nanomed 8:3715–3728
Zhu X, Su M, Tang S, Wang L, Liang X, Meng F, Hong Y, Xu Z (2012) Synthesis of thiolated chitosan and preparation nanoparticles with sodium alginate for ocular drug delivery. Mol Vis 18:1973–1982
Zimmer AK, Kreuter J, Robinson JR (1991) Studies on the transport pathway of PBCA nanoparticles in ocular tissues. J Microencapsul. 8(4):497–504
Zimmer AK, Zerbe H, Kreuter J (1994) Evaluation of pilocarpine-loaded albumin particles as drug delivery systems for controlled delivery in the eye. *In vitro* and *in vivo* characterisation. J Control Release 32:57–70

Chapter 10
Other Advances in Ocular Drug Delivery

Lunawati Bennett

Abstract Ocular drug delivery has many challenges due to the physiology of the eye and the many natural barriers that most drugs need to encounter to permeate the intended tissues. Although traditional eye drops are invasive and convenient, they are inefficient for several ocular diseases due to their low ocular bioavailability and difficulty in delivering a drug to the posterior segment of the eye. Procedures such as implants or frequent intravitreal injections are invasive and challenging; however, this challenges present unique opportunities for innovative drug delivery approaches. New approaches to ocular drug delivery are aimed at: overcoming the disadvantages of existing therapies, overcoming short ocular contact time, increasing low bioavailability drugs to permeate tissues better, limiting dosing frequency, and reducing the invasiveness of some methodologies. This chapter discusses other significant advances in ophthalmic drug delivery such as gene therapy, iontophoresis, sonophoresis, and use of microneedle, hydrogels, and punctual plug delivery systems. A method for restoring light sensing in using retinal prosthetics, optogenetics, and chemical photoswitches are also discussed.

Keywords Advances in ocular delivery · Iontophoresis · Microneedles · Gene therapy

Introduction

Gene therapy is aimed to deliver intracellular genetic material to block a dysfunctional gene or to deliver a gene for a therapeutic purpose (Kompella et al. 2013). Gene therapy was first tried and showed the promising result to treat serious disease Leber Congenital Amaurosis (LCA), an autosomal recessive disease due to a mutation in at least 15 genes that cause blindness in newborn and children.

L. Bennett (✉)
School of Pharmacy, Union University, 1050 Union University Drive,
Jackson, TN 38305, USA
e-mail: llbennett@uu.edu

R.T. Addo (ed.), *Ocular Drug Delivery: Advances, Challenges and Applications*,
DOI 10.1007/978-3-319-47691-9_10

The promising result in clinical trials for LCA have prompted the study of other diseases using gene therapy such as glaucoma, age-related macular degeneration (AMD), retinitis pigmentosa, choroideremia, and others (Solinis et al. 2015).

Hydrogels are cross-linked chemically and physically three-dimensional structures that can control drug release in response to a stimulus. The gels are mostly liquid at room temperature and solid at body temperature. When implanted in the subconjunctival space, it can also sustain the release of drugs delivered to the posterior segment.

The hollow microneedle is an experimental device using injection for intrascleral and suprachoroidal spaces delivery of drugs. In animals, the insertion site was no longer visible 1 h after the injection and the eye appears indistinguishable from a naïve eye. In the clinical setting, microneedle injection of drugs into the suprachoroidal space (SCS) can be used as outpatient surgery procedure to treat diseases such as AMD, diabetic retinopathy and other eye disorders, similar to intravitreal injection, but less invasive (Patel et al. 2011, 2012).

Animal models used for studying ocular drug delivery are usually rabbits, dogs or monkeys because of their close structural and content proximity to human eyes. Animals usually have lower vitreous volumes than humans, but higher vitreous gel contents such as in rhesus monkeys about 60 % and in rabbits almost 100 % of gel, respectively. The anterior chamber and lens are comparatively larger in rabbits, but the vitreous volume is larger in the monkeys. In humans, the vitreous volume is about 4–5 ml with about 40–80 % of gel depending on the age. The monkey model usually provides the best predictive value, while other mammalian models may be selected on the basis of ethical concerns and cost. However, mathematical modeling and simulation should be used to predict human physiological response using data collected from several different animal species. Rats or rodents are not optimal models for scaling exposure or response relationships to humans. Summary of recent ocular delivery in development is listed by Chen (2015).

Gene Therapy

Currently, there are 33 clinical trials reported to have been approved, in progress, or have been completed in the database of Gene Therapy Clinical Trials Worldwide. Viral vector using Adeno-Associated Viruses (AAVs) were used in 22 of the 33 trials. Gene silencing using short interfering riboxy nucleic acid (RNA) (siRNA) was used in 5 of the trials, confirming that gene replacement therapy has a promising clinical approach in the treatment of ocular disease. To treat ocular diseases with gene therapy, there are three parameters that need to be considered for successful results: (a) the administration route, (b) the delivery system, and (c) the use of specific promoter elements (Gaudana et al. 2012).

Administration Routes

The administration route of gene therapy can be through: topical instillation, periocular route, intracameral injection, intravitreal injection, subretinal injection, or suprachoroidal injection. Each administration route shows advantages and disadvantages. The selection of route should be based on the targeted cells and the characteristics of the vector (Solinis et al. 2015). The scheme of different routes of administration is shown in Fig. 10.1.

The topical route is the easiest non-invasive method of gene delivery; however, penetration to the cornea and conjunctival epithelial to reach the posterior segment is limited to the size of the gene that can be delivered; therefore, gene therapy of topical route is usually limited for the treatment of an anterior segment of the eye.

The periocular route includes the administration of drugs to peribulbar, retrobulbar, posterior juxta scleral, sub-tenon, and subconjunctival injection. Subconjunctival injection is the most studied of gene delivery. The drug can penetrate to the anterior and posterior segment, but also can cause systemic absorption and adverse effects. Large particle and nucleic acid drugs are retained for a long time in this part.

Intracameral injection has been used to control intraocular pressure (IOP) with success in the equine model due to the gene injected into the corneal endothelial

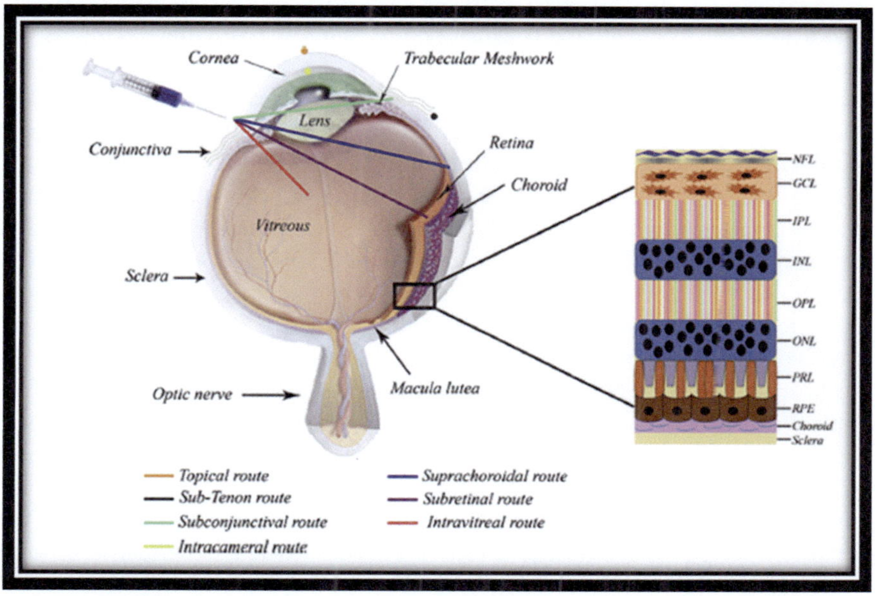

Fig. 10.1 Scheme of different ocular administration routes. *NFL* nerve fiber layer; *GCL* ganglion cell layers; *IPL* inner plexiform layer; *INL* inner nuclear layer; *OPL* outer plexiform layer; *ONL* outer nuclear layer; *PRL* photo receptor layer; *RPE* retinal pigment epithelia with permission from Solinis et al. (2015)

cells and trabecular meshwork produced stable protein expression (Shen et al. 2015; Gaudana et al. 2010).

Several works using intravitreal or subretinal injections showed a promising result for treatment of retinal degenerations of ganglion cells and photoreceptors; however, this is an invasive method that can cause ocular damage such as lesions in retinal pigment epithelial (RPE), hemorrhages, retinal tears, and retinal detachment.

Suprachoroidal, below the sclera and above choroid, administration of gene therapy using non-viral plasmid deoxynucleic acid (DNA) and electrical application have been successfully done in rat eyes without interfering with optical pathway (Kompella et al. 2010; Gaudana et al. 2012).

Delivery System

The ideal vectors to deliver gene should be: (a) they can take the gene efficiently into the target tissue, (b) vectors should carry large amount of genes to induce ocular improvement and also to stay long time in the eye compartment, (c) the vector should be well tolerated and not causing adverse effects such as inflammation, or activating immune response or toxic to the eye. Vectors commonly being researched are categorized into: (1) viral vectors and (2) non-viral vectors (Manning et al. 2002)

Viral vectors continue to be the gene delivery of choice to treat ocular diseases. These vectors include: (a) adenovirus vectors (AVs), (b) adeno-associated viral vectors (AAVs), (c) lentiviral vectors (LVs), and (d) retrovirus vector (RVs). Adenovirus is double stranded DNA vectors that are able to effectively transfect dividing and non-dividing cell without integrating their nucleic acid into the host genomes, thus avoiding the risk of mutagenesis but still large enough genes can be incorporated. However, AVs can induce an immune response and the duration of gene expression is relatively short. In a laboratory study, AVs are useful to carry interleukin (IL) 10 and nerve growth factor gene to improve allograft survival in corneal transplants rats. AVs have been designed to carry genes involved in retinitis pigmentosa, macular degeneration, diabetic neuropathies, and to reprogram fibroblasts to retinal ganglion-like cells in rats (Mohan et al. 2012; Solinis et al. 2015).

Adeno-associated viruses are non-pathogenic single strand DNA vectors that are able to transduce slow or non-dividing cells and to provide long-term gene transcription and expression for up to 6 years. AAVs may be recombined with capsid protein to increase efficacy and specificity, leading to the discovery of recombinant AAVs by some laboratory. The first number in the recombinant vectors denotes the serotype, while the second number corresponds to capsid (AAV n1/n2). For example, AAV2/8 was shown to effectively treat X-linked retinoschisis in mice model by slowing degeneration of the retina and repairing a retinal structure. As recombinants, AAV2/5, AAV2/8, AAV2/9 and AAV2/6 have been shown to be more effective than the original AAVs. Some researchers had a patent application for use in deficient herpes simplex virus-1 (HSV) that can incorporate growth factors, neurotrophins, cytokines, and drugs. In an attempt to deliver nucleic acid which encodes pigment epithelium-derived factor (PEDF), recombinant AAVs has been successfully done in animal (Mohan et al. 2012; Surace and Auricchio 2008).

Lentiviruses are single-stranded RNA vector, are the most suitable vector due to their high and stable expression, ability to integrate cells, self-inactivating, and do not cause severe adverse effects. LVs have been studied on corneal endothelial cells of mice, sheep, and human to transduce genes related to retinal dystrophies, angiostatic proteins endostatin, and angiostatin to treat AMD, macular degeneration, and diabetic retinopathy, and other neovascular diseases (Staout and Appukuttan 2006).

Retrovirus vector carries RNA viruses but can cause oncogenicity. The use of retroviruses for gene therapy has been drastically reduced due to the development of T-cell leukemia in 4 and death of one patient enrolled in the clinical trials for X-linked recessive trials (Mohan et al. 2012).

Non-viral vectors contain cationic compound that binds electrostatically the genetic material and forms stable complex. The non-viral vector can be lipid based or polymer based vectors. Cationic lipids are positively charged amphiphilic molecules that form complexes with negatively charged nucleic acids. Modification of the composition and chemical structures of the lipids such as incorporating protamine sulfate, polyethylene glycol (PEG), or arginine-glycine-aspartate peptides, dextran, chitosan, hyaluronic acid and others are intended to increase nucleic acid protection from degradation by enzymes, to improve cell internalization, or to decrease the trafficking inside the cells, so gene delivery system is more effective. Many novel biodegradable polymers such as poly ethylene imine (PEI), polyesters, chitosan, hyaluronic acid, albumin, poly-L-Lysine (PLL), poly (glycolic) acid (PGA) dendrimers are among polymer based vector that have been studied in improving gene delivery system (Gaudana et al. 2012; Han et al. 2012; Tamboli et al. 2011; Solinis et al. 2015).

Short interfering nucleic acid (siNA) can modulate the gene expression and are designed as anti-angiogenic targeting genes responsible for angiogenesis such as vascular endothelial growth factor (VEGF) type 1, 2, 3 or connective tissue growth factor that cause glaucoma or macular degeneration. Several recent patents in gene therapies include: (a) application of lentiviral vectors to transduce mitotically active and inactive cells, (b) use of recombinant AAV to deliver anti-angiogenic factor, (c) use of mammalian gene CACNA1F encoding for mutated retinal calcium channel that causes congenital stationary night blindness, and (d) use of novel electroporation device to deliver DNA to specific site of the eye (Gaudana et al. 2012; Vargeese et al. 2005).

Promoter Elements

Promoters play an important role by engineering vector to specific tissue targeting and by enabling vector to turn on and off gene expression, and by responding to specific environmental signals. Several tissue-specific promoters for corneal and retinal cells such as keratin 12 (epithelial specific), keratocan (keratocyte specific), cone and rod homeobox, rhodopsin, rod opsin have been tested successfully in the laboratories.

Gene therapy has been tested in several ocular diseases such as LCA, X-linked retinoschisis (XLRS), Stargardt disease, choroideremia, retinitis pigmentosa, AMD, a disease of the cornea, and glaucoma. Glaucoma affects about 3 % of people over 40 years worldwide by causing an increase in intraocular pressure (IOP) and if not treated on time can cause blindness. Gene therapy offered improvement in lowering IOP by inhibiting beta 2 adrenergic receptor via the use of siRNA and SYL040012 (a double strand oligonucleotides) as shown in clinical trials in 24 healthy volunteers (Solinis et al. 2015).

XLRS is retinal degenerative disease due to the mutation in gene encoding retinoschisis. This progressive disease can cause severe loss. Gene therapy with non-viral vectors and retinoschisis 1 (RS1) plasmid in an animal model showed promising results in slowing the progression of the disease (Solinis et al. 2015).

Stargardt disease is inherited juvenile macular degeneration that usually occurs in younger than 20 years old. The gene encodes for this disease is due to mutation of ABCA4. After the gene therapy, significant correction and functional structure are observed. There is a clinical trial that is currently underway using gene therapy (Solinis et al. 2015).

Choroideremia is an X-linked chorioretinal dystrophy that affects male. Mutation in CHM gene is associated with this disease. Successful gene therapy using AAV. REP1 in male patients showed improvement in the visual function and slowing down of the degeneration rate (Solinis et al. 2015).

Retinitis pigmentosa is the most common retinal degeneration responsible for the loss of vision of 1 in 4000 people worldwide. Defects in 60 genes include the mutation in inherited autosomal dominants, autosomal recessive, X-linked, or rhodopsin gene (RHO) have been identified. Suppression of mutant genes was successfully obtained using siRNA with AAV delivery (Solinis et al. 2015).

LCA is an autosomal recessive disease that causes marked impairment of visual acuity at birth. There is no successful treatment for LCA. Gene therapy has been conducted successfully in 4 clinical trials using lentiviral vector and AAVs to treat LCA (Solinis et al. 2015) (Fig. 10.2).

Iontophoresis

Iontophoresis is a non-invasive technique in which low electric current is applied to enhance ionized drug penetration into tissue. Iontophoresis increases drug penetration primarily through two mechanisms: electromigration (orderly movement of ions in the presence of an electrical current) and electro-osmosis (a convective solvent flow in the anode to cathode direction that occurs under a physiological condition when the epithelium is negatively charged). Neutral non-ionized molecules are more effectively delivered when anodal iontophoresis was used

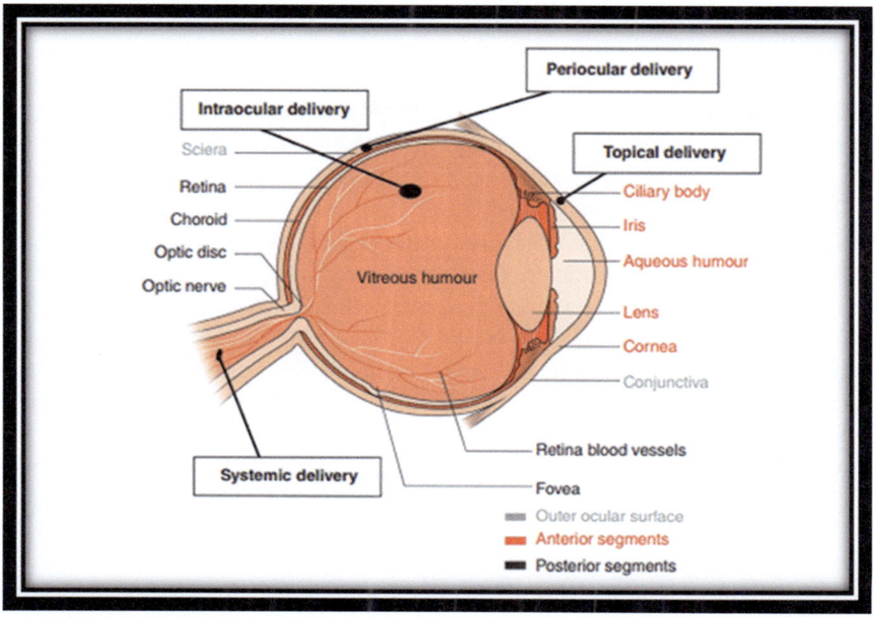

Fig. 10.2 Intraocular drug delivery bypasses anatomic and dynamic barriers of the posterior segment. The placement of a therapeutic substance directly into the vitreous (intravitreal injection) or the space between the retina and RPE (subretinal injection) though invasive, can achieve the highest intraocular bioavailability by bypassing several anatomical and dynamic barriers of the posterior segment. With permission from Shen et al. (2015)

(Kompella et al. 2010). This showed the importance of electro-osmosis on the transport of uncharged drugs. The negatively charged molecules are more effectively delivered when cathodal iontophoresis was used showing electromigration contributes to the corneal penetration of macromolecules (Horwath-Winter et al. 2005).

The drug is applied using an electrode carrying the same charge as the drug. An electrode with the opposite charge is placed elsewhere in the body to complete the circuit. Ionized drug substance through the anode (for a positively charged drug) or the cathode (for a negatively charged drug), or neutral molecules is delivered based on electro-osmosis or solute-associated fluid transport which under the action of the electrical current will penetrate into tissue (Vaka et al. 2008).

Iontophoresis may be an attractive option for patients who are not responsive to eye drop therapy. Iontophoresis technology is used to direct drugs to the target sites, due to ease of use, safe and limited systemic adverse effects. Several preclinical trials have been done successfully to deliver antibiotics such as ciprofloxacin or gentamicin (treatment of pseudomonas keratitis), antifungals, or non-steroidal anti-inflammatory drugs (NSAIDs) or steroids (dexamethasone or methylprednisolone to treat inflammation), antisense oligonucleotides (treatment of angiogenesis), carboplatin (treatment of retinoblastoma), and methotrexate (treatment of intraocular lymphoma) to anterior or posterior segments using trans-corneal or transscleral iontophoresis. In general, transscleral iontophoresis is more effective for drug delivery to the posterior segment, while trans-corneal is more effective for anterior segment. The increase in the electrical current density and the duration of application enhance the drug penetration. In general currents for up to 3 mA for 20 min or 1.5 mA for up to 40 min are well tolerated without causing adverse effects or obvious blurred vision or pain (Souza et al. 2013).

Recently, iontophoresis method has been used to deliver nanoparticles. Positively charged nanoparticles showed better penetration into the inner tissue than negatively charged nanoparticles. Positive charged interact better with negatively charge cornea and conjunctiva mucosa (Souza et al. 2013).

The Visulex Company use transscleral iontophoresis with the shape of a lens to deliver a drug to posterior segment using 2 compartments: drug ion to be delivered and a counter ion to precipitate the drug. When the current is applied, the counter ion such as calcium and the drug ions penetrate the sclera, they then precipitate and form a depot. This system decreases the need for frequent application. Besides calcium, other ions such as Sn^{+2}, Fe^{+2}, Fe^{+3}, Mn^{+2}, Zn^{+2}, NH^{+4}, organic anion, chelating agents have also been tried in an attempt to decrease the clearance of the drugs from the desired site, so sustained delivery of drug can be achieved.

The Eyegate II Company makes ocular iontophoresis deliver dexamethasone phosphate to decrease sign and symptoms of inflammation in the anterior and posterior segments. The EyeGate II system design for trans-corneal drug delivery consists of an ocular applicator, syringe, adaptor for transferring the drug product from reservoirs to the applicator and generator to provide consistent current to the electrode as shown in Fig. 10.3. Figure 10.4 shows the change in the eye of rabbit after reverse iontophoresis (RI) (Sun et al. 2015).

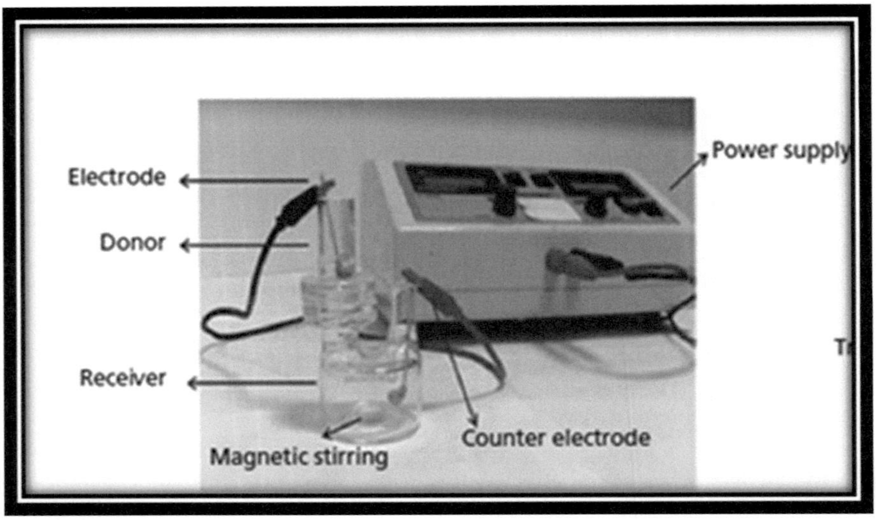

Fig. 10.3 Iontophoresis with electrode, generator, and syringe, with permission from Souza et al. (2013)

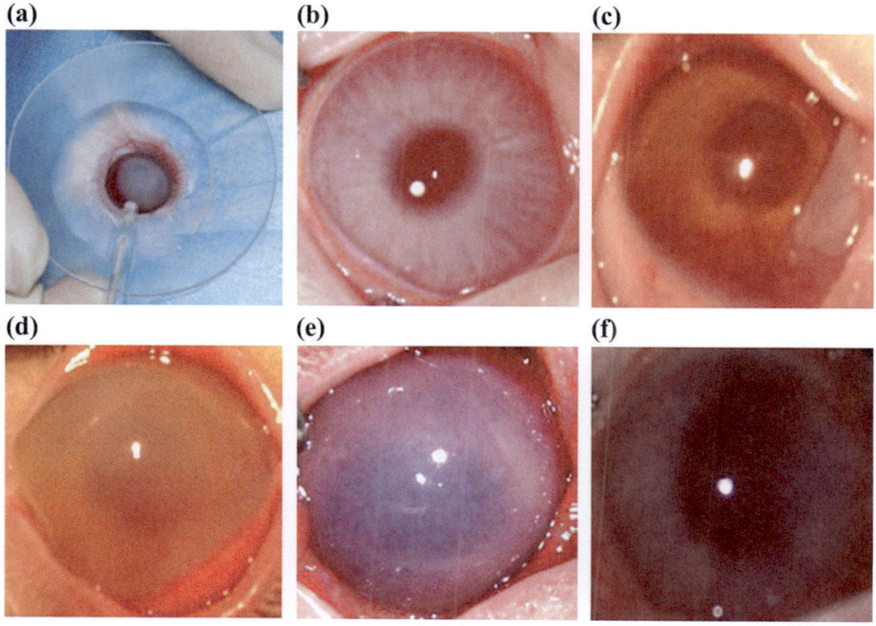

Fig. 10.4 Anterior segment of the eye. **a** Reverse iontophoresis (RI) cathode electrode and the eyecup. The cup is placed on the eye cornea, and Ag/AgCl electrode is inserted into the eyecup to provide the cathode current. **b** The normal rabbit eye with a smooth and transparent cornea, and a transparent anterior segment in the control group. **c** Immediate image after anterior segment iron foreign body model. The anterior segment was brown, and oozing around the pupil was observed. **d** 24 h after anterior iron foreign, there were conjunctival hyperemia, corneal edema, large bullous keratopathy and rust colored pigmentations and anterior segment was brown. **e** Six days with 0 mQA current RI. **f** Six days after 0.4 mA current RI. With permission from Sun et al. 2015 Open access from CCAL

Sonophoresis

Sonophoresis use ultrasound at low frequency (20–100 kHz), medium frequency (400–800 kHz) or high frequency (7000–16000 kHz). Low-frequency ultrasound was more efficient in increasing penetration of hydrophilic ocular drugs. High ultrasound was used to enhance the intrascleral delivery of peptide drugs by enhancing the permeation with no damage to retinal tissue even when using high-frequency ultrasounds (Souza et al. 2013) (Fig. 10.5).

Microneedles

The microneedle is invented to inject a drug into a certain area of the posterior segment of the eye such as suprachoroidal space (SCS). AMD with or without diabetic neuropathy (DN) can cause blindness. Currently, drugs used to treat this

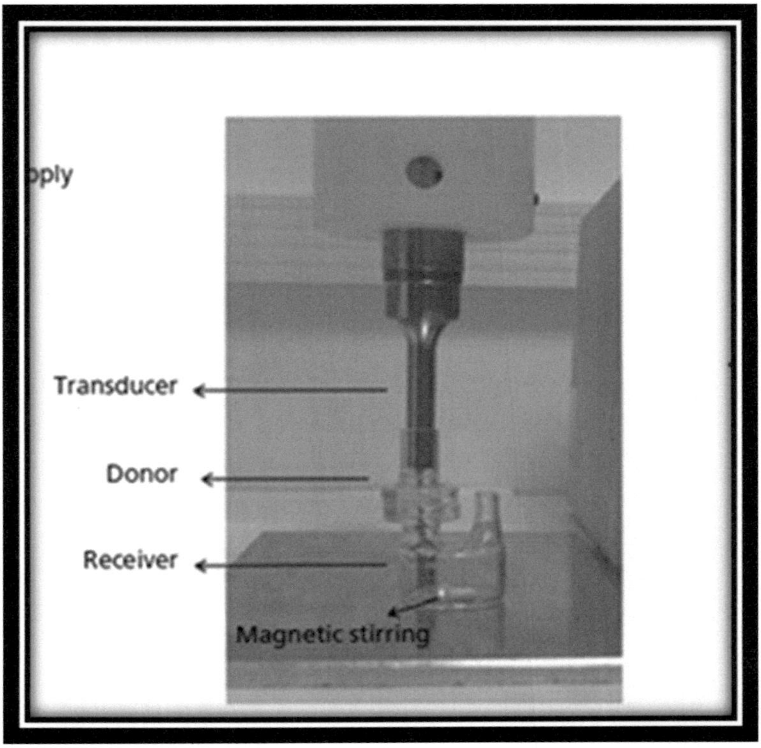

Fig. 10.5 Sonophoresis with an electrode, generator, and syringe. Picture with permission from Souza et al. (2013)

disease are delivered to choroid or retina using a intravitreal method by injecting a liquid formulation into the vitreous or by placing extended release implants in the vitreous (Kim et al. 2007; Lee et al. 2010). This method doesn't provide an effective drug concentration in the intended area of the choroid and retina (Patel et al. 2012).

To target a drug to treat AMD or DN more specifically closer to the choroid and retina, the SCS which is a space located between sclera and choroid can provide higher drug levels in the target tissue. For this reason, the SCS provide a promising site for administration of a drug to treat other disease related posterior segment also. In the past, to access the SCS site, either the surgical procedure or use of a long cannula or hypodermic needles is needed, but this technique is cumbersome. To improve this cumbersome techniques, injection into the SCS using a hollow glass microneedle was invented (Patel et al. 2011, 2012). Figure 10.6 showed the comparison of a microneedle for SCS injection to a 30-gauge hypodermic needle.

Targeting SCS provide accurate dosing and decrease exposure of drugs to non-targeted tissues. In the study using rabbit cadaver, Patel et al. demonstrated that clearance of molecules and particles injected into the SCS occurs at different rates.

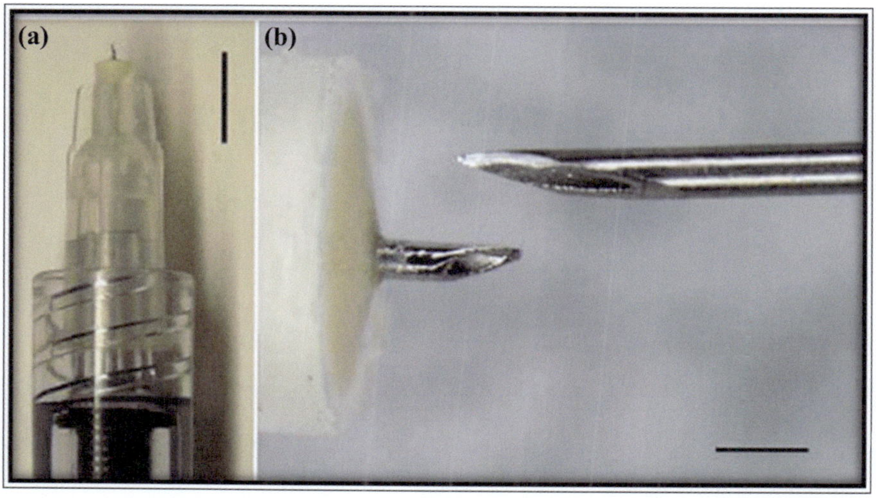

Fig. 10.6 Microneedle for SCS injection. Low magnification view of a microneedle at the end of a syringe (**a**) and high magnification comparison of a microneedle (*left*) to the tip of a 30 gauge hypodermic needle (*right*). *Scale bar* 5 mm (**a**) and 500 μm (**b**). With permission from Patel et al. (2012)

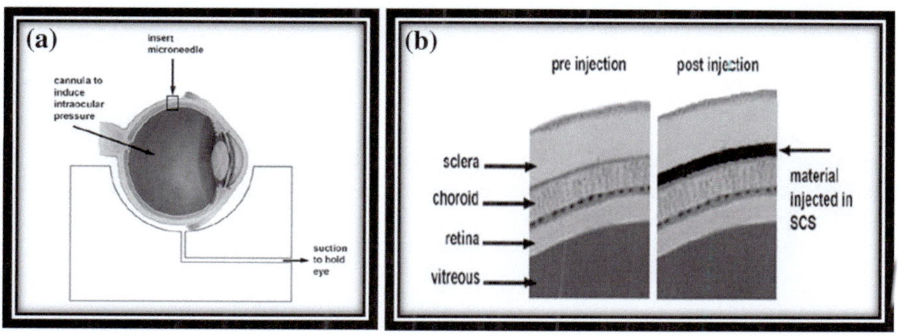

Fig. 10.7 **a** Schematic diagram of the experimental setup used to study suprachoroidal injection in which rabbit eye was placed on a custom-made mold with a channel through which suction was applied to hold the eye in the place. A cannula inserted through the optic nerve of the eye allowed application of intraocular pressure. The microneedle was inserted into the sclera using the custom made insertion device. (**b**) A magnified view of the boxed area in (**a**) shows an idealized schematic of the anatomy of the periocular tissues near the insertion site before and after the proposed suprachoroidal. With permission from Patel et al. (2011)

Figure 10.7 showed the schematic diagram for the microneedle injection. Figure 10.8 Showed rabbit eye tissue after microneedle injection.

Small soluble molecules or macromolecules exhibited a short period of time while nano or microparticles remain in the SCS for months with no sign of clearance. Suprachoroidal injection of drug loaded particles enabled the drug to be

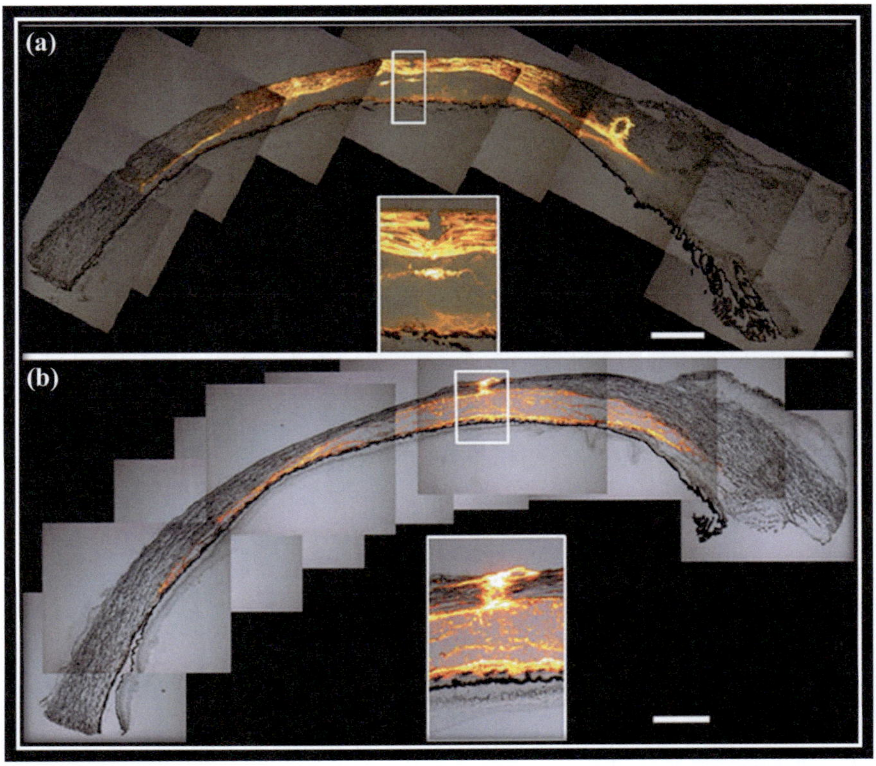

Fig. 10.8 Each of particle size on particle distribution in the eye. Collaged fluorescence microscopy images of tissue cryosections show the delivery of **a** (*top*) 20 nm particles and **b** (*below*) 1000 nm particles into the suprachoroidal space of pig eyes ex vivo. These images show that 20 nm particles spread into the suprachoroidal space and within the sclera. However, the 10,000 nm particles are primarily in the suprachoroidal space. The insertion sites are magnified in the *insets*. *Scale bar* 500 µm. With permission from Patel et al. (2011)

delivered to chorioretinal tissue for as short as half day or as long as several months based on particle degradation kinetics on how the drug is formulated. The use of microneedle causes less trauma to the eye because the needle tract is much shorter and narrower than other approaches and can provide a straightforward way to access the SCS.

Hydrogels

Hydrogels are hydrophilic system composed of polymers which form a three-dimensional (3D) network that water infiltrates and enable the drug to diffuse in and out of the gel. Hydrogels can be used as drug reservoirs and are more resistant to removal from the eye through blinking because of their viscoelastic properties (Eljarrat-Binstock et al. 2008). A hydrogel is liquid at pH 6 and at room temperature, but undergo gelation at pH 7.4, the natural pH of the tear film. The hydrogel can deliver approximately 95 % of the drug and the drug can stay more than 8 h. in the eye. Different polymers can be blended to improve the hydrogel characteristics. Poloxamers, a thermo-sensitive in situ forming gel has weak mechanical strength; however, by blended poloxamers with polyacrylic acid, alginate, and chitosan, this combination of gel stayed at longer time on the eye surface. Sodium alginate is another polymer that is converted into the gel in the presence of calcium which is present in the lachrymal fluid. Sodium alginate with hydroxyl propyl methyl cellulose (HPMC) was formulated with moxifloxacin to improve sustained release of the drug from less than 4 h (in regular solution form) to more than 10 h in hydrogel form (Yang et al. 2012).

A blend of chitosan with other thermosensitive polymers such as poly-N-Isopropyl acrylamide showed increase permeability of drug such as timolol maleate for the treatment of glaucoma (Kompella et al. 2013). Commercial products known to treat glaucoma such as Timoptic-XE® is based on ion-activated gel that uses gellan gum as the ion-sensitive agent (made by Merck), while Zirgan (Bausch & Lomb) contains polyacrylic acid and ganciclovir as the active compound to treat herpetic keratitis (Yang et al. 2012).

A hydrogel can also be used for replacing nylon suture. Corneal wounds from surgical procedures (transplants, incisions for cataract removal, intraocular lens implantation), ulcer infections and traumatic injuries (laceration, perforation) are usually repaired using nylon suture. Multiple sutures are often needed to realign the edge of damaged tissue in effort to restore the structural integrity of the cornea. However, nylon suture has several disadvantages such as: trauma to corneal tissue, increase in corneal scarring, loose or broken suture, irregular astigmatism, and suturing requires technical skill that can vary widely from surgeon to surgeon. Therefore, an adhesive like hydrogel to replace or supplement sutures in the repair of corneal wound is a preferable technique. Hydrogel network can be formed using: (a) photocrosslinking, in which upon exposure to visible light, the acylated-modified dendritic macromolecules crosslinks to form a hydrogel or (b) nucleophile-electrophile reaction in which the crosslinking can occur at 37 °C under neutral aqueous solutions and chemoselective reaction (Grinstaff 2007). Figure 10.9 shows the hydrogel network. Figure 10.10 shows application of photo-crosslink hydrogel to the laceration followed by irradiation that sealed the wound in rabbit eye. Figure 10.11 shows comparison of the change between suture and hydrogel.

Punctual Plug Drug Delivery System (PPDS)

Punctual plugs (Fig. 10.12) have been used for more than 20 years for symptomatic relief of dry eye syndrome and glaucoma. Punctual plugs used for drug delivery can be made from polymers in a variety of shapes and sizes. The punctual plugs are

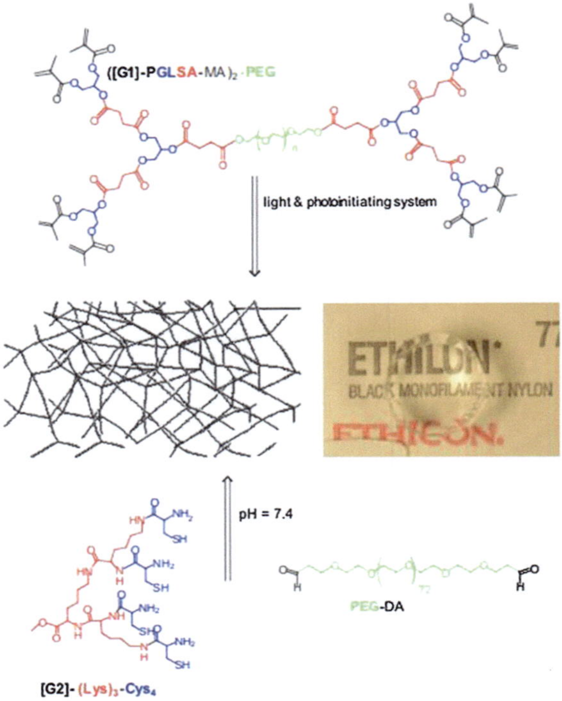

Fig. 10.9 (*Top*) Schematic of photocrosslinking reaction. (*Bottom*) Nucleophile-electrophile reaction. Picture with permission from Grinstaff (2007)

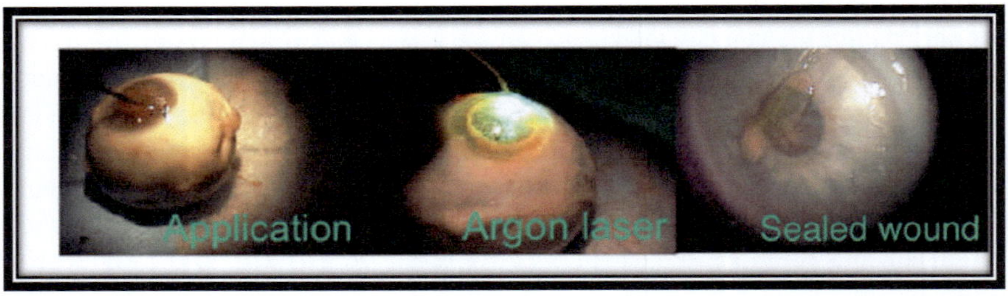

Fig. 10.10 Photographs of the closure procedure for a 4.1 mm full corneal laceration (*left*). Placement of the adhesive solution to the wound (*middle*). Photocrosslinking of the solution to form the adhesive hydrogel seal the corneal laceration (*right*). Picture with permission from Grinstaff (2007)

composed of: (a) cylindrical body containing the drug compound, (b) an optional outer shell made up of material impermeable to the drug and the tear fluid, (c) an optional cap material containing pores, and (d) an optional unit to retain the punctual plug over long period. The bottom end is tapered and narrower to allow

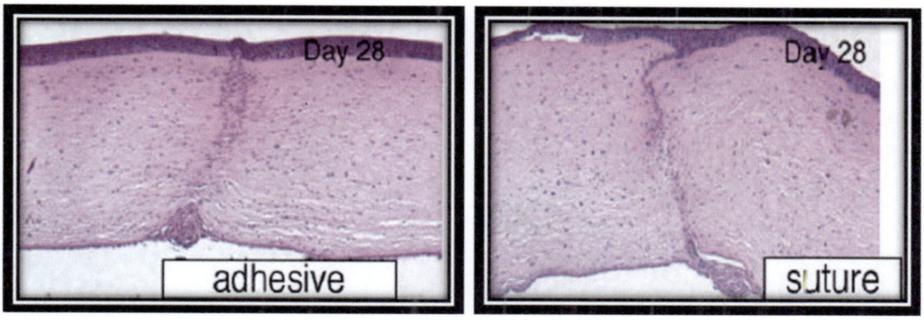

Fig. 10.11 Stains of the repair chicken cornea after treatment with hydrogel dendrimer adhesive and after suture. Picture with permission from Grinstaff (2007)

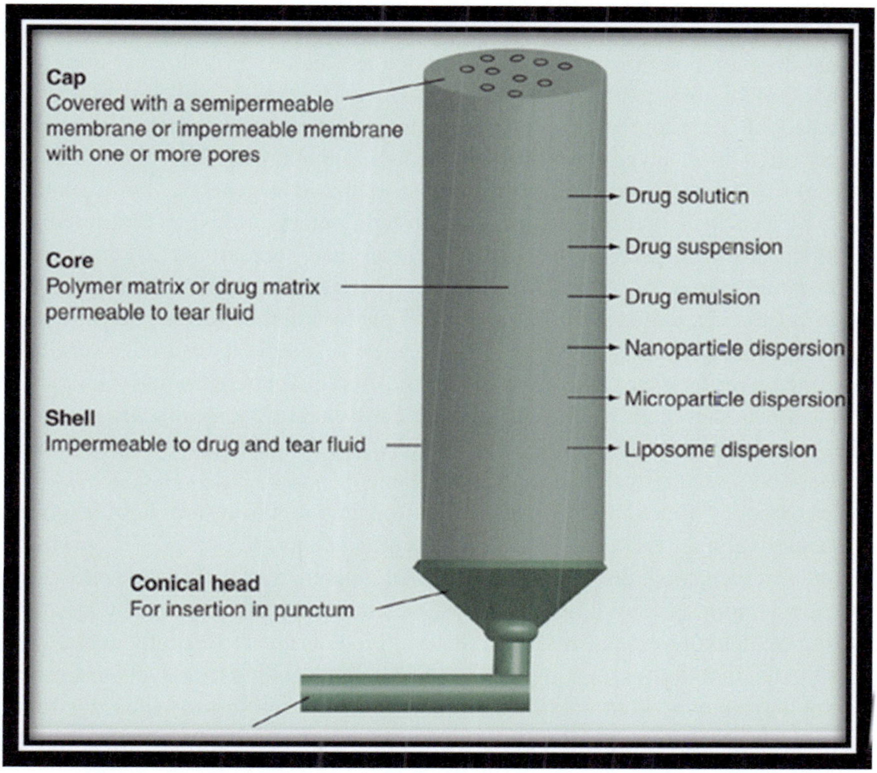

Fig. 10.12 Punctual plug delivery system with various components and drug-loading methods. The scheme captures a variety of technologies that are under development and is not intended to represent any technology completely. With permission from Kompella et al. (2010)

easy insertion into the punctum the head portion is exposed to tear film. The drug is released from a punctual plug by diffusion from the polymeric core to the tear fluid. The drug can be loaded into the central polymeric core as solution, suspension, microemulsions, nanoparticles, microparticles or liposomes with or without an additional polymer matrix (Kompella et al. 2010).

PPDS can be made from silicone, Teflon, hydroxyl ethyl methacrylate (HEM), polycaprolactone (PCL) or polydioxanone. The drug can stay for up to 180 days, after which the plug can be removed. Latanoprost-PPDS has been shown to decrease IOP in glaucoma patients with some adverse effects such as eye itchiness, eye irritation, increased lacrimation and ocular discomfort (Kompella et al. 2010).

Retinal Prosthetics, Optogenetics, Chemical Photoswitches

Retinal degeneration is inherited disorders that are directly or indirectly due to the death of rod photoreceptor cells. Many of these orders are called retinitis pigmentosa (RP), a name that reflects the common disease phenotype of exposed pigmentation from invading retinal pigmented epithelium (RPE) cells. There are over 200 sources of gene defects that cause retina to degenerate, but eventually many RP patients suffer total visual loss (no light perception, NLP) or are reduced to impair bare light perception (BLP) (Nirenberg and Pandarinath 2012).

There are strategies for restoring light-driven signaling with clinical potential: retinal prosthetics, optogenetics, or chemical photoswitches. Argus II retinal prosthetic technology use a stimulation arrayed with 60 platinum electrodes surgically positioned over the inner surface of retina (the epiretinal configuration), coupled to an embedded scleral electronics package via transscleral ribbon cable and driven by a head-mounted camera. This system permits significant recovery of positional, motion, and structural percepts in many patients (Zrenner 2013) (Fig. 10.13).

Optogenetics uses viral delivery of type 1 opsin genes to restore light responses in the survival neurons. Genes for various type 1 opsins are packed with promoter or enhancer elements into AAV viral gene that delivers by subretinal or intravitreal injection as non-replicating, long expression gene in animal neurons. By hijacking existing trafficking mechanisms, the translated proteins are successfully delivered to the cell membrane thus making the residual retinal cells become photosensitive again. Optogenetics could provide the ultimate in high-resolution vision due to the long persistent gene expression in animal models; however, optogenetics remains challenging to be implemented in human eyes with large volumes, complexity of the disease, and physical barriers for the viral penetration (Busskamp et al. 2012; Marc et al. 2014).

Chemical switchers use small molecules such as acrylamide-azobenzene-quaternary ammonium (AAQ) or better with diethylamino-azobenzene quaternary ammonium (DENAQ) that modulates neuronal electrical activity. Chemical photo switches act rapidly after a single intravitreal injection with no sign of toxicity. Targeting high-density ganglion cells with chemical photoswitches may rapidly

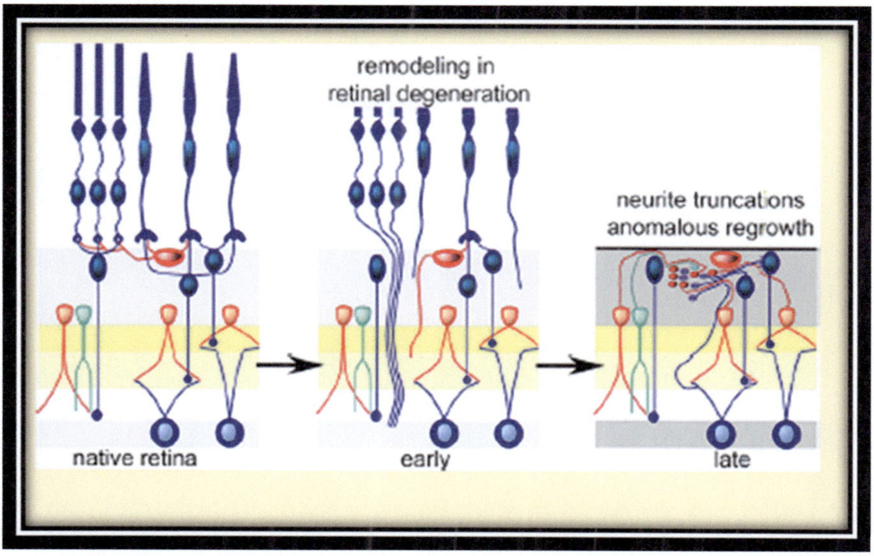

Fig. 10.13 Cellular composition of the retina and the phases or remodeling associated with retinal degeneration. With permission from Marc et al. (2014)

generate a high-resolution vision, and develop photo switches with longer physiological half-lives or delivery strategies providing long-term depots (Marc et al. 2014; Toschitsky et al. 2014).

Summary

Gene therapy has advanced since its introduction almost 20 years ago. Currently, there are over 100,000 articles in gene-based therapy for the treatment of many diseases in some clinical trials and laboratory animals. Although ocular gene therapy is still in its early stages of development, the potential of gene-based intervention for treatment of blindness, corneal diseases, abnormal wound healing, are progressing. Ocular drug delivery using iontophoresis improved electrical current density and application time which increase bioavailability of macromolecular drugs or drugs formulated as nanoparticles to the anterior and posterior segments of the eye. Transscleral iontophoresis is more suitable for posterior segment whereas application through the cornea is more effective for the anterior. Sonophoresis used ultrasound intensity and frequency improved drug permeability to corneal and scleral. Punctual plug using latanoprost to treat ocular hypertension and open angle glaucoma are in phase 2 clinical trials.

Some of the new drug delivery systems in the developmental stage are listed in Table 10.1

Table 10.1 Summary of drug delivery technologies in development (Chen 2015)

Technology	Description	Company	Phase	Indication
Punctual plug	Drug loaded polymeric, biodegradable, reside in puncta to release drug over time, non-invasive, focus on anterior diseases	Mati Therapeutics Ocular Therapeutix	2 3	Glaucoma Post-operation inflammation, pain, allergic conjunctivitis
Implants				
Polymerics	Biodegradable, implanted in subconjunctival or invitreal to release drug over time, invasive, focus on posterior diseases	pSivida, Allergan	2 2 1 3	Posterior uveitis, glaucoma, atrophy, retinitis pigmentosa, macula-off retinal detachment
Refillable	Refillable, non-degradable, subconjunctival to release drug over time, initial invasive, refillable less invasive, focus on posterior diseases	Genentech Replenish	1 1	Wet AMD, DME
Encapsulated cells	Reservoir, invasive, non-biodegradable, focus on posterior diseases	Neurotech Pharmaceuticals	2, 2, 1	Retinitis pigmentosa, macular telangiectasia type 2, atrophy, wet AMD, glaucoma, neuropathy, optic nerve stroke
Topical inserts	Soft elastomers, non-degradable, rest under eyelids over sclera to release drug over time, non-invasive, focus on anterior segment	ForSight Vision5, Amorphex Therapeutics	2, Preclinical	Glaucoma
Contact lens	Drug containing soft contact lens, non-degradable, contact lenses with drug reservoir that drug release over time, non-invasive, focus on anterior segment	Various research groups	Preclinical	Glaucoma
Iontophoresis	Wearable electrical device to drive drug into ocular tissue, non-invasive, focus on anterior and posterior diseases	EyeGate Pharma	3	Anterior uveitis, dry eye, cataract surgery

(continued)

Table 10.1 (continued)

Technology	Description	Company	Phase	Indication
Gene therapy	Viral vector-based delivery, injected to deliver genetic material, invasive, focus on posterior diseases	Spark Therapeutics, Avalance Biotechnologies, Genzyme, Oxford BioMedica	1, 2 2 1 1	Choroideremia, Wet AMD, LCA, Stargardt disease, Usher syndrome
Particulate systems	Polymeric/liposomal drug crystal, biodegradable, topical or injectable, focus on anterior and posterior diseases	Kala Pharmaceuticals	3 2	Ocular inflammation, post cataract surgery, dry eye

Abbreviation LCA Leber congenital amaurosis; *DME* Diabetic macular edema; *AMD* Age-related macular degeneration

References

Busskamp V, Picaud S, Sahel JA, Roska B (2012) Optogenic therapy for retinitis pigmentosa. Gene Ther 19:169–175
Chen H (2015) Recent development in ocular drug delivery. J Drug Target 23:597–604
Eljarrat-Binstock E, Orucov F, Frucht-Pery J et al (2008) Methylprednisolone delivery to the back of the eye using hydrogel iontophoresis. J Ocul Pharmacol Ther 24(3):344–350
Gaudana R, Ananthula HK, Parenky A, Mitra AK (2010) Ocular drug delivery. AAPS J 12:348–360
Gaudana RJ, Gokulgandhi MR, Boddu SHS, Mitra AK (2012) Recent overview of ocular patents. Recent Pat Drug Deliv Formul 6(2):95–106
Grinstaff MW (2007) Designing hydrogel adhesives for corneal wound repair. Biomaterials 28 (35). doi:10.1016/j.biomaterials.2007.08.041
Han Z, Conley M, Makkia R et al (2012) Comparative analysis of DNA nanoparticles and AAVs for ocular gene delivery. PLoS ONE 7:e521819
Horwath-Winter J, Schmut O, Haller-Scholar EM et al (2005) Iodide iontophoresis as a treatment for dry eye syndrome. Br J Opthamol 89(1):40–44
Kim SH, Galban CJ, Lutz RJ et al (2007) Assessment of subconjunctival and intrascleral drug delivery to the posterior segment using dynamic contrast-enhanced magnetic resonance imaging. Invest Opthalalmol Vis Sci 48:808–814
Kompella UB, Kadam RS, Lee VHL (2010) Recent advances in ophthalmic drug delivery. Ther Deliv 1(3):435–456
Kompella UB, Amrite AC, Ravi RP, Durazo SA (2013) Nanomedicines for back of the eye drug delivery, gene delivery, and imaging. Prog Retin Eye Res 36:172–198
Lee S, Hughes P, Ross A, Robinson M (2010) Biodegradable implants for sustained drug release in the eye. Pharm Res 27:2043–2053
Manning WCJ, Dwarki VJ, Rendahl K, et al (2002) Use of recombinant gene delivery vectors for treating or preventing disease of the eye. US20020194630
Marc R, Pfeiffer R, Jones B (2014) Retinal prosthetics, optogenetics, and chemical photoswitches. ACS Chem Neurosci 5:895–901
Mohan RR, Tovey JCK, Sharma A, Tandon A (2012) Gene therapy in the cornea: 2005-present. Prog Retin Eye Res 31(1):43–64
Nirenberg S, Pandarinath C (2012) Retinal prosthetic strategy with the capacity to restore normal vision. Proc Natl Acad Sci USA 109:15012–15017
Patel SR, Lin AS, Edehauser HF, Prausnitz MR (2011) Suprachoroidal drug delivery to the back of the eye using hollow microneedles. Pharm Res 28:166–176
Patel SR, Berezovsky DE, McCarey BE et al (2012) Targeted administration into the suprachoroidal space using microneedle for drug delivery to the posterior segment of the eye. Invest Opthalalmol Vis Sci 53(8):4433–4441
Shen HH, Chan EC, Lee JH et al (2015) Nanocarriers for treatment of ocular neovascularization in the back of the eye: new vehicles for ophthalmic drug delivery. Nanomedicine (Lond) 10:2093–2107
Solinis MA, Rodriquez ADP, Apaolaza PS, Rodriquez-Gascon A (2015) Treatment of ocular disorders by gene therapy. Eur J Pharm Biopharm. doi:10.1016/j.ejpb.2014.12.022
Souza JG, Dias K, Pereira TA et al (2013) Topical delivery of ocular therapeutics: carrier systems and physical methods. J Pharm Pharmacol 66:507–530
Staout JT, Appukuttan B (2006) Lentiviral vector-mediated gene transfer and uses thereof. US20060062765
Sun S, Diao H, Zhao F et al (2015) Extraction of iron from the rabbit anterior chamber with reverse iontophoresis. J Opthalmol. doi:10.1155/2015/425438
Surace EM, Auricchio A (2008) Versatility of AAV vectors for retinal gene transfer. Vision Res 48:353–359

Tamboli V, Mishra GP, Mitra A (2011) Polymeric vectors for ocular gene delivery. Ther Deliv 2 (4):523–536

Tochitsky I, Polosukhina A, Degtyar VE et al (2014) Restoring visual function to blind mice with a photoswitch that exploits electrophysiological remodeling of retinal ganglion cells. Neuron 81:800–813

Vaka SR, Sammeta SM, Day LB, Murthy SN (2008) Transcorneal iontophoresis for delivery of ciprofloxacin hydrochloride. Curr Eye Res 33(8):661–667

Vargeese C, Wang WM, Chen T et al (2005) Polycationic composition for cellular delivery of polynucleotides. US200050222064

Yang H, Tyagi P, Kadam RS et al (2012) Hybrid dendrimers hydrogel/PLGA nanoparticle platform sustains drug delivery for one week and anti-glaucoma effects for four days following one-time topical administration. ACS Nano 6:7595–7606

Zrenner E (2013) Fighting blindness with microelectronica. Sci Transl Med 5:210–216

Retraction Note to: In Vitro and in Vivo Evaluation of Ocular Drugs and Delivery Systems

Ruhi V. Ubale and Richard T. Addo

Retraction Note to: Chapter 7 in: R.T. Addo (ed.), *Ocular Drug Delivery: Advances, Challenges and Applications*, https://doi.org/10.1007/978-3-319-47691-9_7

The authors have retracted this chapter [1] because of overlap with a previously published article by Shafaie, S. et al. [2]. All chapter authors agree with this retraction.

[1] Ubale R.V., Addo R.T. (2016) In Vitro and in Vivo Evaluation of Ocular Drugs and Delivery Systems. In: Addo R.T. (ed) Ocular Drug Delivery: Advances, Challenges and Applications. Springer, Cham.

[2] Shafaie, S., Hutter, V., Cook, M. T., Brown, M. B., & Chau, D. Y. S. (2016). In Vitro Cell Models for Ophthalmic Drug Development Applications. BioResearch Open Access, 5(1), 94–108. http://doi.org/10.1089/biores.2016.0008.

The retracted online version of this chapter can be found at https://doi.org/10.1007/978-3-319-47691-9_7

R.T. Addo (ed.), *Ocular Drug Delivery: Advances, Challenges and Applications*,
https://doi.org/10.1007/978-3-319-47691-9_11

MIX
Papier aus verantwortungsvollen Quellen
Paper from responsible sources
FSC® C105338

If you have any concerns about our products,
you can contact us on
ProductSafety@springernature.com

In case Publisher is established outside the EU,
the EU authorized representative is:
Springer Nature Customer Service Center GmbH
Europaplatz 3, 69115 Heidelberg, Germany

Printed by Libri Plureos GmbH
in Hamburg, Germany